EOS Optical Technologies Conference 2019

EPJ Web of Conferences Volume 215 (2019)

Munich, Germany
24 - 26 June 2019

Editors:

O. Fahnle
A. Vasdekis

ISBN: 978-1-5108-9425-9

Printed from e-media with permission by:

Curran Associates, Inc.
57 Morehouse Lane
Red Hook, NY 12571

Some format issues inherent in the e-media version may also appear in this print version.

For additional information, please contact EDP Sciences – Web of Conferences
at the address below.

EDP Sciences – Web of Conferences
17, Avenue du Hoggar
Parc d'Activité de Courtabœuf
BP 112
F-91944 Les Ulis Cedex A
France

Phone: +33 (0) 1 69 18 75 75
Fax: +33 (0) 1 69 28 84 91

contact-edps@webofconferences.org

Additional copies of this publication are available from:

Curran Associates, Inc.
57 Morehouse Lane
Red Hook, NY 12571 USA
Phone: 845-758-0400
Fax: 845-758-2633
Email: curran@proceedings.com
Web: www.proceedings.com

TABLE OF CONTENTS

EPJ Web of Conferences **215**, 01001 (2019) https://doi.org/10.1051/epjconf/201921501001
EOS Optical Technologies

Coherent Ray Tracing Simulation Of Non-Imaging Laser Beam Shaping With Multi-Aperture Elements

Raoul Kirner[1,*], *Wilfried* Noell[1], *Toralf* Scharf[2], and *Reinhard* Voelkel[1]

[1]SUSS MicroOptics SA, CH-2068 Hauterive, Switzerland
[2]NAM, École Polytechnique Fédérale de Lausanne (EPFL), CH-1015 Lausanne, Switzerland

Abstract. The application of laser light sources for illumination tasks like in mask aligner lithography relies on non-imaging optical systems with multi-aperture elements for beam shaping. When simulating such systems, the traditional approach is to separate the beam-shaping part (incoherent simulation) from dealing with coherence properties of the illuminating laser light source (diffraction theory with statistical treatment). We present an approach using Gaussian beam decomposition to include coherence simulation into ray tracing, combining these two parts, to get a complete picture in one simulation. We discuss source definition for such simulations, and verify our assumptions on a well-known system. We then apply our approach to an imaging beam shaping setup with microoptical multi-aperture elements. We compare the simulation to measurements of a similar beam-shaping setup with a 193 nm continuous-wave laser in a mask-aligner configuration.

1 Introduction

To improve the resolution in proximity mask aligner lithography, decreasing the exposure wavelength λ is the only way if a certain exposure gap g is to be maintained as resolution is given by [**1**]:

$$Resolution \sim \sqrt{g \cdot \lambda}\,. \qquad (1)$$

This requires light sources emitting at shorter wavelengths than the classically used Hg high-pressure arc lamps, like excimer lasers or TOPTICA's continuous-wave 193 nm diode laser light source [2].

Mask aligner lithography relies on uniform illumination of the photomask, containing the structural information which is to be transmitted to a photoresist-coated substrate. Uniform illumination, both in intensity and over a finite angular spectrum, is ensured using an imaging homogenizer setup featuring a microoptical Koehler integrator as mixing element [3,4], see Fig. 1.

Replacing the partially coherent Hg arc lamps in mask aligners with a coherent light source requires the addition of means to the optical system to deal with the effects of coherence. Otherwise Fabry-Perot responses from all interfaces in the optical path would be observable in the resist image. This was achieved by introducing rotating optical diffusers into the illumination system [2]. While illumination systems relying on rotating diffusers can be modelled by incoherent ray tracing with good results, the investigation of coherence effects in the illumination path requires coherent simulation of the system [5].

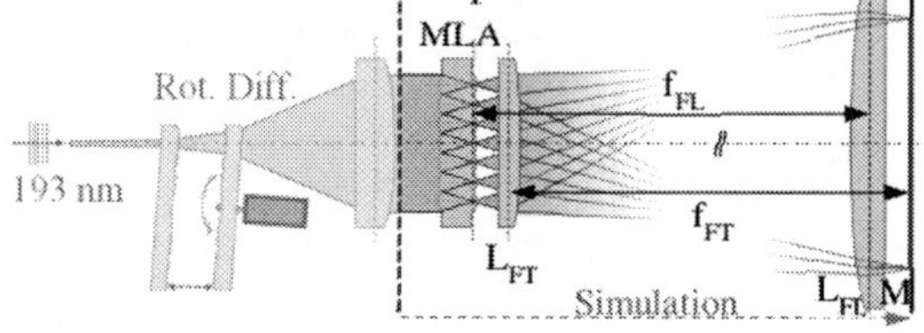

Fig. 1. Mask aligner illumination system for a coherent 193 nm light source. A diffuser setup featuring a rotating diffuser controls coherence in the mask plane M. An imaging beam homogenizer with microoptical Koehler integrator, consisting of microlens arrays (MLA), and Fourier and Field lenses L_{FT} and L_{FL} ensure uniform and telecentric illumination in the mask plane.

Coherent simulation of such systems is demanding: one has to deal with some thousands of microoptical channels in combination with macroscopic length-scales for the overall illumination system and a diffuse yet coherent source. We present an approach using Gaussian beam decomposition (GBD) for modelling coherence in ray tracing simulation as an approach to combine these opposing requirements.

2 Coherent ray tracing of diffuse sources

This work was carried out using PHOTON ENGIEERING's FRED ray tracing software, where GBD is implemented.

GBD relies on the approximation of a wavefront by an equally spaced array of Gaussian beamlets. Each one representing the local average intensity and (relative) phase of the wavefront, and propagating in the direction of the local surface normal. Each beamlet consists of a main or base ray and secondary rays which model beamlet waist and divergence. To model the coherent response of an optical system in good approximation these beamlets are coherently summed up on a detector plane after being propagated through the system [6,7].

To test the validity of this approach for simulating illumination systems with optical diffusers, we consider a simple test system: a single lens focusing the wavefront after an ideal optical diffuser illuminated by a plane wave of diameter D onto a detector plane in focal distance f behind the lens, compare Fig. 3.

The illuminated diffuser is approximated by a GBD plane wave source, where a random phase between 0 and 2π is added to each beamlet. On the detector a speckle pattern is visible whose lateral average speckle size σ_{sim}

* Corresponding author: raoul.kirner@suss.com

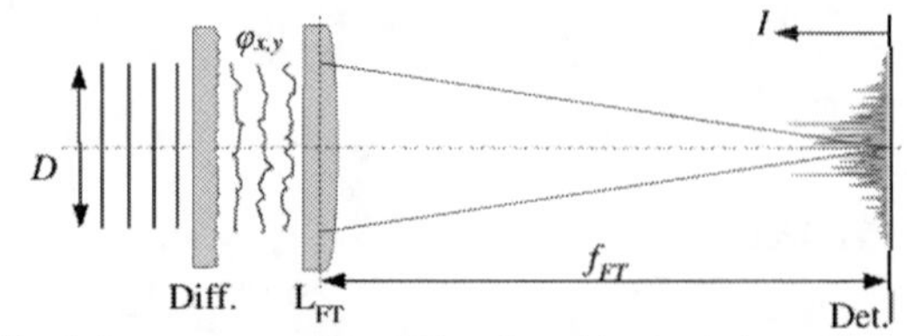

Fig. 3. Layout to test the validity of speckle simulation with GBD. A plane wave of diameter D is shone onto a diffuser where their phase profile $\varphi_{x,y}$ is locally changed. A Fourier lens focuses the wavefront onto a detector in focal distance behind it. There the intensity distribution is captured.

can be found as the width of the autocorrelation peak of said field [8]. We compare this simulated speckle size to the theoretically expected speckle size of an ideal coherent source after an ideal diffuser in focal distance behind an ideal lens, the diffraction limited spot size [9],

$$\sigma_{x,y} \approx 1.22\frac{\lambda f}{D}. \quad (2)$$

Figure 3 shows this comparison, and furthermore illustrates the robustness of this approach. The approach delivers constant results for a large variety of source sampling rates and source diameters. For σ_{sim} the autocorrelation peak is measured at full width half maximum (FWHM), containing approximately 68% of the total energy, compared to the expected diffraction limited spot size, containing roughly 84% of the energy. This leaves a discrepancy of about 4% to the found mean value for the speckle size ratio S which can be contributed to the noise floor of the simulation.

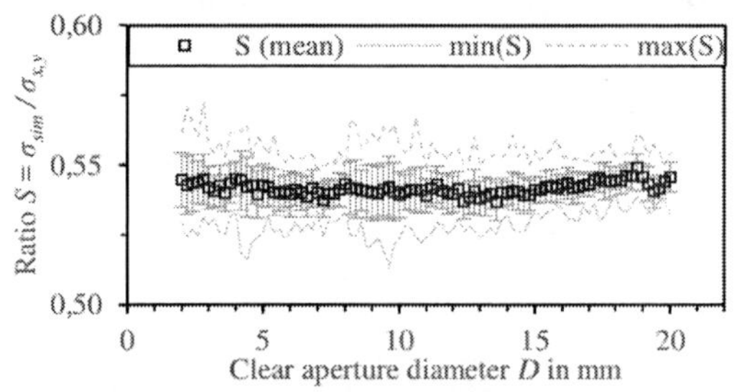

Fig. 2. Ratio S of simulated FWHM speckle size σ_{sim} to theoretical lateral speckle $\sigma_{x,y}$ size as function of beam diameter D. Error bars show the standard deviation of variable source sampling (10…250 λ beamlet spacing). Dashed lines show local minimum and maximum values of S.

3 Simulation of a multi-aperture system

To coherently simulate an imaging homogenizer setup the path before the MLA is replaced by a randomized GBD source as described in the previous part, compare Fig. 1.

The source sampling has been carefully chosen to allow for successful implementation: deletion of beamlets has to be avoided by ensuring that all rays of each beamlet always pass the same surfaces (one condition of GBD). This is critical when modelling multi-aperture elements like MLA to maintain correct optical power transmission.

Fig. 4 a) shows the simulated intensity distribution for one random source. Temporal averaging, as implemented with a rotating diffuser in combination with the integrating nature of photoresist as sensor, can be

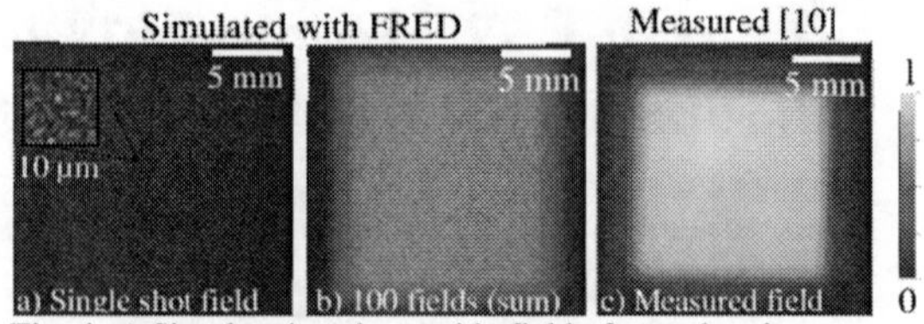

Fig. 4. a) Simulated static speckle field of a random input source after the beam-shaping system as shown in Fig. 1. b) Superposition of 100 random speckle fields in the mask plane. c) Comparison to a measurement of the same setup, taken with a Silicon photodiode with 300 µm square aperture.

modelled by incoherently superimposing multiple simulation runs (static images) with different random initial optical path length distributions added to the source beamlets, as shown in Fig. 4 b). For comparison, Fig. 4 c) shows a measurement of this illumination system.

4 Conclusion

We have found conditions under which Gaussian beam decomposition is a valid method to simulate beam-shaping systems with microoptical multi-aperture elements. We discussed how to simulate a randomized source in the ray tracing software FRED. We utilized this approach to investigate beam-shaping systems featuring multi-aperture elements for mask aligner lithography in the DUV with a 193 nm CW laser light source including diffusers for coherence control.

The authors would like to thank Andreas Vetter and TOPTICA PHOTONICS for the close cooperation.

References

1. P. Rai-Choudhury, ed., *Handbook of micro-lithography, micromachining, and microfabrication, Volume 1* (SPIE Press, 1997)
2. R. Kirner, A. Vetter, D. Opalevs, C. Gilfert, M. Scholz, P. Leisching, T. Scharf, W. Noell, C. Rockstuhl, R. Voelkel, Opt. Express **26**(2), 730 (2018)
3. R. Völkel, U. Vogler, A. Bich, P. Pernet, K. J.Weible, M. Hornung, R. Zoberbier, E. Cullmann, L. Stürzebecher, T. Harzendorf, U. D. Zeitner, Opt. Express **18**, 20968–20978 (2010)
4. F. M. Dickey, ed., *Laser beam shaping: Theory and Techniques 2nd Ed.*, chap. 10 (CRC Press, 2017)
5. R. Völkel, K. J. Weible: Proc. SPIE **7102**, 1020J (2008)
6. J. Arnaud, Appl. Opt. **24**, 538-543 (1985)
7. A. W. Greynolds, Proc. SPIE **0679** (1986)
8. J. W. Goodman, *Statistical Optics 2nd Ed.*, chap. 7.7 (Wiley, 2015)
9. M. Sjödahl, L. R. Benckert, Appl. Opt. **33**(31), 7461-7471 (1994)
10. R. Kirner, A. Vetter, D. Opalevs, M. Scholz, P. Leisching, T. Scharf, W. Noell, C. Rockstuhl, R. Voelkel, Proc. SPIE **10587**, 105870W (2018)

EPJ Web of Conferences **215**, 01002 (2019) https://doi.org/10.1051/epjconf/201921501002
EOS Optical Technologies

Development of a Cost-Efficient Computer Controlled Optical Surfacing Process for Correcting Aspheric Lenses using Tool Influence Function based Dwelltime Optimization

Pei Liang Low[1,*], *Wilhelmus A. C. M.* Messelink[1], and *Rene* Weber[1]

[1]Edmund Optics Singapore Pte. Ltd., 18 Woodlands Loop #04-00, Singapore, Singapore

Abstract. A Computer Controlled Optical Surfacing (CCOS) system has been developed for correcting form errors on aspheric surfaces. Experiments were carried out to find the correlation between different polishing parameters and polishing metrics such as removal rate, uniformity etc. Based on established polishing parameters, polishing process is developed to correct surface errors on planar, spherical and aspheric surfaces. A convolution model between TIF and dwell times was developed to simulate and solve for correction polishing. Surface accuracies of peak-to-valley (PV) 141 nm and root-mean-squared (RMS) 22 nm has been achieved for planar surface. For aspheric surface, current accuracy of 662 nm PV and of 115 nm RMS is achieved with further development ongoing.

1 Introduction

The global market is moving towards tighter and tighter specifications on aspheres, while at the same time pushing for a reduction in unit price. This trend has been ongoing for several years and Edmund Optics understands customers need for mid-range, price-sensitive aspheres, which requires an alternative corrective technology than the current, costly, state-of-the-art.

To improve the accuracy of a CNC polished asphere, it is typically 'finished' on a much more expensive, corrective polishing platform (such as Magneto-Rheological Finishing [1], Precessed Bonnet Polishing [2] or Ion-Beam Figuring [3]). These so-called Computer Controlled Optical Surfacing (CCOS) platforms are commercially available and in widespread use around the world, and are being installed more and more in low-cost countries. Using these platforms on the lower-end of their capability scale will therefore become a very competitive and uneconomical use of our machine park. This is summarized in the capability versus cost matrix in in Table 1, where the red colour denotes an area of low economic value.

To fill the highlighted gap we have started the in-house development of a cost-efficient CCOS platform to serve the moderately precise asphere market. We achieve this by using proven, of-the-shelf components, flexible processes and existing industry standards.

2 Approach

In CCOS [4] the surface form error and the removal characteristics of the finishing process are measured and quantified, after which a computer algorithm calculates the process parameters to be used to correct the surface.

Table 1. Capability versus cost matrix

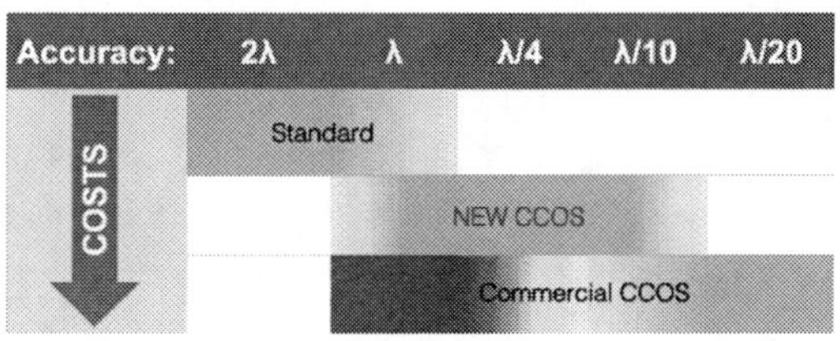

The process is characterised by the "Tool Influence Function", the removal per unit of time as a function of the position within the working area of the process. If the process is linear in time, then the removal at a point on the surface, $R(\vec{x})$, depends on where the process is held on the surface, $\vec{x}_p$, how long it is held there, the dwell time $d(\vec{x}_p)$, and the influence function of the process, $f(\vec{u})$,

$$R(\vec{x}) = d(\vec{x}_p) \cdot f\left(\vec{x} - \vec{x}_p\right) \quad (1)$$

If the process is moved over the whole surface of the workpiece along a path $P : \vec{x}_p(t)$, which is a smooth curve, then the combined removal becomes

$$R(\vec{x}) = d \otimes f \quad (2)$$

the convolution of the dwell times with the influence function of the process. The desired removal is equal to the starting surface error $e(\vec{x})$ and a computer algorithm minimizes the predicted residual errors, $r(\vec{x})$,

$$r = e - d \otimes f \quad (3)$$

to find the optimal dwell times along the path.

*e-mail: pllow@edmundoptics.com.sg

https://doi.org/10.1051/epjconf/201921501002

The finishing process is using industry standard polishing slurry (e.g. cerium oxide) and polyurethane polishing pads mounted on a spindle. While the pad is moved over the surface, the orientation of the spindle is held normal to the surface being polished, which requires a kinematic system with five degrees of freedom (three for translation and two for orientation).

3 Results

A planar work piece of 45 mm is uniformly polished over a 25 mm square to evaluate the stability and to estimate the removal rate of the process. Figure 1 shows a 2D slice

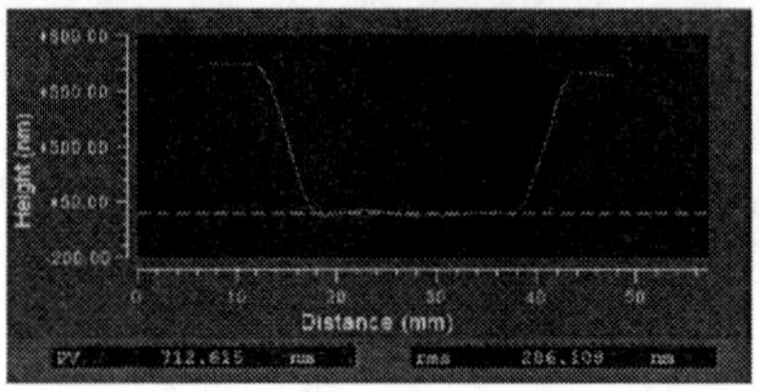

Figure 1. 2D profile through uniform removal square

across the removal area, showing stable removal. The inner area has a uniformity of 10.2 nm RMS. Based on the volume of removal and total polishing time, an estimated removal rate is established.

After calibrating the removal, corrective polishing was performed on a planar, 50 mm diameter workpiece with a starting form error of 362 nm PV and 99 nm RMS. The

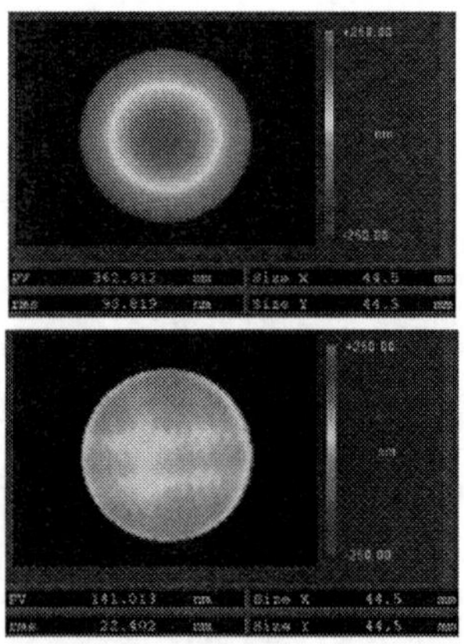

Figure 2. Interferometric measurement of plano before and after correction

resulting form error after corrective polishing is 141 nm PV and 22 nm RMS.

The next experiment involves polishing an aspheric lens using a deconvolution algorithm which takes into the account of the TIF of the polishing pad and solves for the optimized dwell time at each polishing point. An 55 mm diameter aspheric lens is used as the work piece and corrective polishing performance is evaluated within 80% of the work piece diameter. Starting with an error of 1339 nm PV and 251 nm RMS, the corrective polishing is able to bring down the error to 662 nm PV and 115 nm RMS. From the metrology data, it is evident that the work

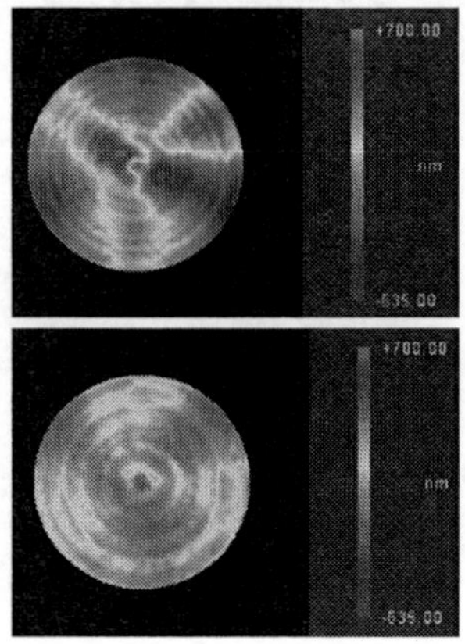

Figure 3. Interferometric measurement of asphere before and after correction

piece has been over-corrected, which results in a slightly inverted error map compared to the input. Hence, this implies that with a better calibrated removal, the correction can be further improved. Moreover, metrology of the aspheric workpiece is performed with a 3D optical profilometer, with limited accuracy. Further development is currently ongoing using measurement data of a Stitching Interferometer, which has superior accuracy.

4 Conclusion

An in-house CCOS system has been designed and built targeting aspheric corrective polishing with an accuracy between $\lambda/2$ and $\lambda/10$ PV. Current results achieved for planar surface is 141 nm PV and 22 nm RMS. Initial trials on aspheric surface results in 662 nm PV and 115 nm RMS with further development ongoing.

References

[1] S.D. Jacobs, D. Golini, Y. Hsu, B.E. Puchebner, D. Strafford, I.V. Prokhorov, E.M. Fess, D. Pietrowski, W.I. Kordonski, *Magnetorheological finishing: a deterministic process for optics manufacturing*, in *International Conference on Optical Fabrication and Testing*, edited by T. Kasai (SPIE, 1995), Vol. 2576 of *Proc. SPIE*, pp. 372–382

[2] D. Walker, D. Brooks, A. King, R. Freeman, R. Morton, G. McCavana, S.W. Kim, Opt. Express **11**, 958 (2003)

[3] S. Zafran, K.A. Kaufman, M.M. Silver, Optical Engineering **21**, 216002 (1982)

[4] R.A. Jones, Appl. Opt. **16**, 218 (1977)

Increasing the Laser-Induced Damage Threshold of Optical Components by Atmospheric Pressure Plasma Surface Finishing

Christoph Gerhard*, *Marco* Stappenbeck, and *Daniel* Tasche

University of Applied Sciences and Arts, Faculty of Natural Sciences and Technology, 37085 Göttingen, Germany

Abstract. In this contribution, a plasma-based approach for finishing optics surfaces is introduced. Experiments were performed on classically manufactured zinc crown glass and sapphire. It is shown that the use of direct dielectric barrier discharge plasma at atmospheric pressure allows the removal of surface-adherent carbonaceous contaminations that were induced by classical manufacturing. Moreover, the use of such plasma leads to a certain decrease in surface roughness. Both effects, surface cleaning and smoothing finally increase the laser-induced damage threshold of optical components.

1 Introduction

In the last decades, the laser-induced damage threshold (LIDT) of optics surfaces has gained in importance due to the rapidly growing number of laser applications and the accompanying need of laser beam shaping and imaging systems. However, the LIDT of optics surfaces is limited, e.g. due to micro cracks, subsurface damages, hydrated silica gel layers and light-absorbing near-surface contaminants. These effects and phenomena result from classical optics manufacturing via grinding and chemical-mechanical polishing. In order to avoid or heal such defects, quite different techniques and methods – mainly post-processing steps – were developed and investigated in the past. In this contribution, a novel and promising plasma-based approach is presented.

2 Materials and methods

In this work, classically lapped and polished plane plates made of zinc crown glass as well as classically ground and super-polished sapphire samples were investigated. The surfaces were finally plasma treated using a direct dielectric barrier discharge (DBD) plasma operated with argon as process gas [1]. The samples were placed within the discharge gap, consequently acting as dielectric barrier and thus as an essential part of the discharge. In doing so, the plasma was ignited directly on the sample surface that was slightly heated up to a temperature of approximately 90°C [2].

The LIDT was determined by the diameter regression technique and the detection of the minimum damage fluence, respectively, as reported in more detail in [3] for zinc crown glass and in [4] for sapphire. LIDT measurements were performed prior to and after plasma treatment. Moreover, the impact of the plasma treatment on surface contamination was investigated via X-ray photoelectron spectroscopy (XPS).

3 Results and discussion

3.1 Zinc crown glass

By the plasma treatment (for 60 seconds), the carbon content on the zinc crown glass surfaces was notably reduced as shown in Figure 1. The presence of carbon can be attributed to carbonaceous contaminations as induced by the manufacturing process, e.g. by wear debris from the lapping agent or the polyurethane polishing pad [3].

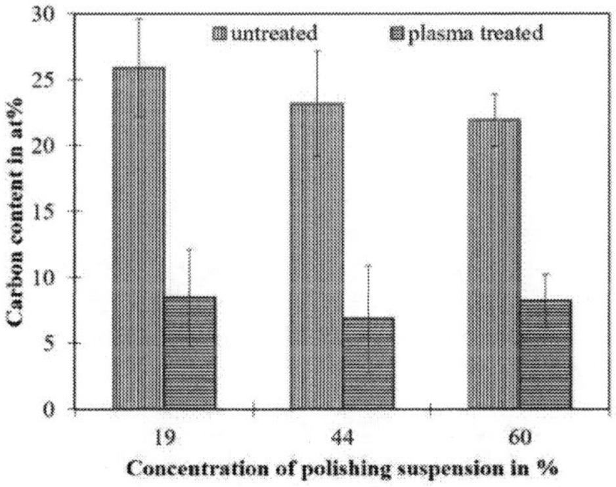

Fig. 1. Carbon content of zinc crown glass surfaces before and after plasma treatment; the samples were polished with different polishing suspension concentrations.

Interestingly, the initial carbon content tends to depend on the concentration of the polishing suspension and the resulting type of interaction between the work piece and the polishing pad surface, respectively. In contrast, the final carbon content after plasma treatment reaches a saturation value. This behaviour can be explained by a plasma-induced removal of carbonaceous contaminants

* Corresponding author: christoph.gerhard@hawk.de

from the surface whereas some residues remain in the subsurface region as for example within micro cracks. However, the plasma cleaning leads to an increase in LIDT as shown in Figure 2.

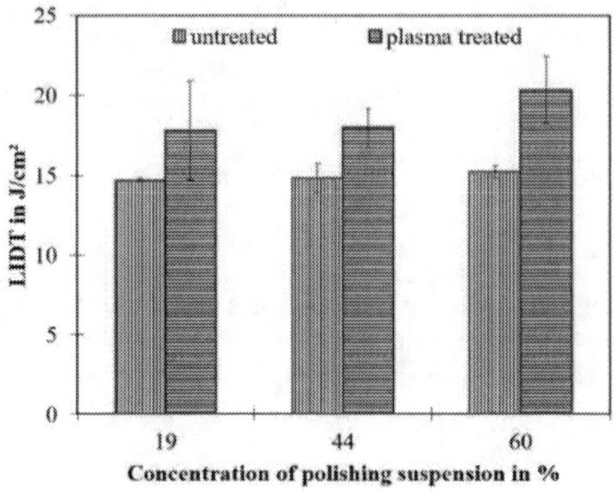

Fig. 2. Laser-induced damage threshold (LIDT) of zinc crown glass surfaces before and after plasma treatment; the samples were polished with different polishing suspension concentrations.

3.2 Sapphire

A plasma-induced removal of carbonaceous contaminants from the surface was also observed in the case of super-polished sapphire as shown in Figure 3.

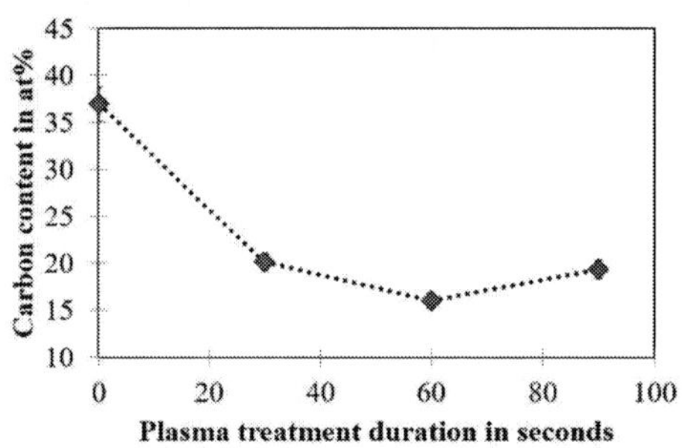

Fig. 3. Carbon content of sapphire surfaces vs. plasma treatment duration.

In addition to XPS measurements, the impact of the plasma treatment on surface roughness was investigated via atomic force microscopy (AFM) measurements where the quadratic mean area roughness Sq of the sample surfaces before and after plasma treatment was detected. As shown in Figure 4, the plasma treatment effected a certain surface smoothing.

This effect was already observed in previous work [1,5] and leads to a reduction of diffusively scattered light at the surface, consequently contributing to an increase in LIDT. The increase in LIDT of the investigated sapphire surface as shown in Figure 5 can thus be explained by both plasma-induced effects, surface cleaning and smoothing. It finally turns out that after a plasma treatment duration of 90 seconds, the LIDT is increased by a factor of 1.7.

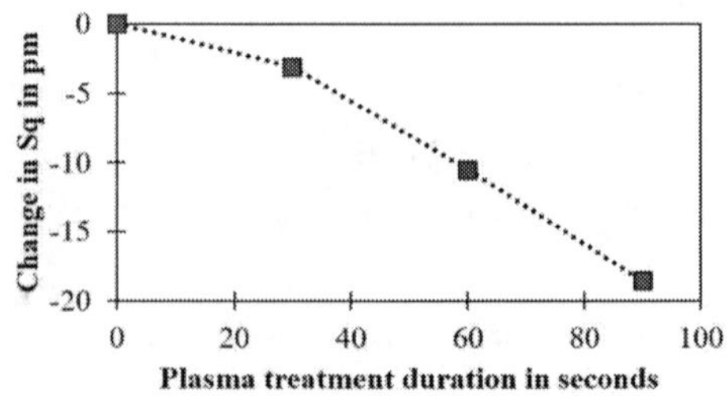

Fig. 4. Change in quadratic mean area roughness Sq of sapphire surfaces vs. plasma treatment duration.

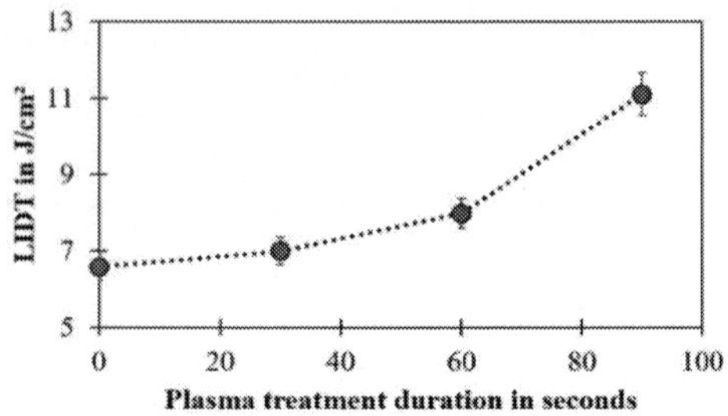

Fig. 5. Laser-induced damage threshold (LIDT) of sapphire surfaces vs. plasma treatment duration.

4 Conclusions

The observed increase in LIDT is of great interest for the manufacture of laser optics and protection windows. The application of a direct DBD plasma represents a cost-efficient approach with a high degree of efficiency. In addition to cleaning, surface smoothing is a further advantageous plasma-induced effect. The short process time makes this technique applicable on industrial scale where the "cold" plasma treatment at temperatures well below the glass softening temperature and the possibility to work at atmospheric pressure are additional benefits.

The authors would like to thank Helene Dyck from IMPEX HighTech GmbH in Rheine as well as Lutz Müller and Robert Köhler from the University of Applied Sciences and Arts in Göttingen for their help during the preparation and analysis of the investigated samples.

References

1. C. Gerhard, T. Weihs, A. Luca, S. Wieneke, W. Viöl, J. Eur. Opt. Soc.-Rapid **8**, 13081 (2013)
2. A. Gredner, C. Gerhard, S. Wieneke, K. Schmidt, W. Viöl, J. Mater. Sci. Eng. B **3**, 346 (2013)
3. C. Gerhard, M. Stappenbeck, Appl. Sci. **8**, 1556 (2018)
4. C. Gerhard, D. Tasche, N. Munser, H. Dyck, Opt. Lett. **42**, 49 (2017)
5. C. Gerhard, S. Roux, S. Brückner, S. Wieneke, W. Viöl, Appl. Opt. **51**, 3847 (2012)

Using A Digital Temperature Sensor To Measure The Temperature Development During A Polishing Process

Benisch Michael[1,*], *Bogner* Werner[1,**], and *Rascher* Rolf[1,***]

[1]Technische Hochschule Deggendorf, Dieter-Görlitz-Platz 1, 94469 Deggendorf

Abstract. The temperature development within a lens is recorded during processing with a digital temperature sensor. The development of the temperature over time shows dependencies on the relative speed and the contact pressure on the polishing wheel. Furthermore, the amount of polishing suspension flowing into the polishing gap may have an influence on the temperature development.

1 Introduction

The temperature, generated by a process contributes to the energy, which is caused by mechanical friction or chemical processes [1]. Therefore, the measurement of the temperature gives a hint on the energy induced in the process. The temperature change (increase or decrease) is measured during a common polishing process.

2 Experimental Setup

The experimenal setup consists of three components:

- Polishing wheel, with a stepwise (0 to 10) adjustable rotation speed between 20 1/min to 900 1/min.
- Polishing suspension (Hastilite PO, uncooled)
- Lens with integrated temperature sensor (SHT31)

Different experiments were carried out, including the measurement of the influence of

- Relative velocity
- Pressure and pressure variations during process
- High pressure and maximum possible velocity

to the temperature that adjusts itself in the lens. The lenses prepared for the experiment can be described as "smart" workpieces. They allow conclusions to be drawn from polishing influences on the resulting temperature. The resulting temperature corresponds to the energy input caused by friction and chemical effects. In a pure mechanical system, the source of heat are the pad-workpiece friction and material removal [1].

2.1 Digital Temperature Sensor

The absolute accuracy of the sensor is ±0.2 K, the repeatability can be chosen manually and is between ±0.2 K to ±0.04 K. Usually their value is ±0.08 K (setting for the experiments). The evaluation takes place via an Arduino Uno R3 microcontroller board, which is connected to the sensor via I2C communication protocol.

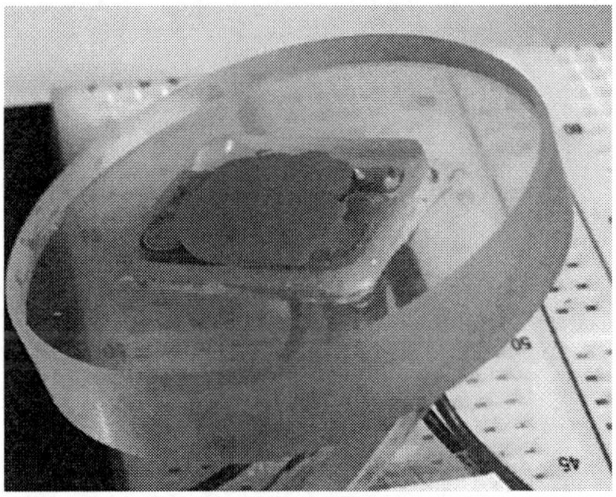

Figure 1: Glass sample with integrated digital temperature sensor. The grey area represents the conductive paste between the glass and the sensor. The dimensions of the sensor are only 2 mm × 2 mm.

The sensor is glued into a milled recess in the lens. The thermal connection to the glass material is made by an electrically non-conductive heat conducting paste. The sensor's optional humidity detection feature was not used during the experiments.

2.2 Proof of principle: Maximum heating

This test was carried out to cause a maximum possible heating of the lens. For this purpose, the workpiece was

*e-mail: michael.benisch@th-deg.de
**e-mail: werner.bogner@th-deg.de
***e-mail: rolf.rascher@th-deg.de

 https://doi.org/10.1051/epjconf/201921501004

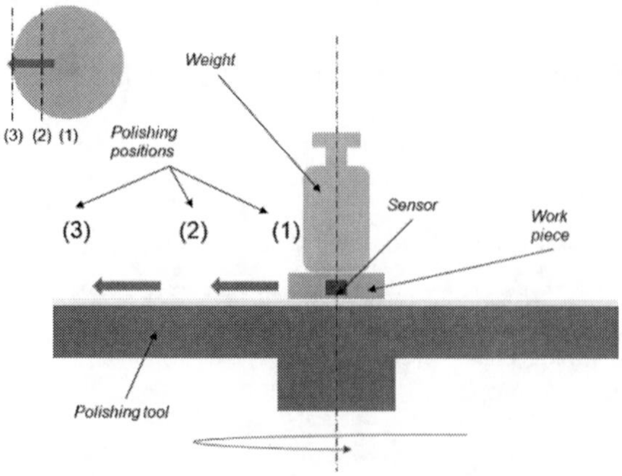

Figure 2: Arrangement of the experiment. The polishing at various points changes the relative speed between tool and workpiece. The weight acts as a force from above on the workpiece equipped with a sensor.

polished at maximum rotation speed 900 1/min and with a high contact pressure at the edge of the polishing wheel in order to achieve a maximum relative velocity between workpiece and tool. The pressing force was introduced by a weight of 2 kg and a manual guidance of the workpiece.

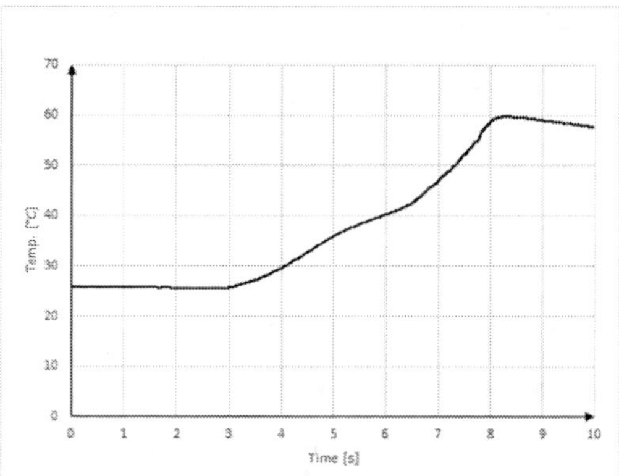

Figure 3: Temperature development of the glass lens shown in figure 1 with approximately constant high pressure and maximum adjustable speed of the polishing wheel. The highest reached value is almost 60 °C. At this point, the experiment was stopped to prevent damage to the polishing wheel. Immediately after finishing the polishing process, a cooling of the workpiece can be observed.

The measurement shows, that the lens can reach a temperature of almost 60 °C under the conditions of a high speed and high pressure process without cooling. It also shows, that cooling starts instantly after ending the processing.

The next experiments will focus on machining at variable speeds and pressures under process conditions (integration of a supply and cooling system for the polishing slurry). The part of mechanical and chemical interactions will be investigated during processing. Especially chemical processes are accelerated or slowed down by the influence of temperature. This presumed influence is to be proven.

The impact of an insufficient supply of polishing suspension is also considered. In addition the division of temperature, which occurs between workpiece, tool and slurry [2] will also be considered by the integration of further sensors close to the process.

References

[1] Evans, C. J.; Paul, E.; Dornfeld, D.; Lucca, D. A.; Byrne, G.; Tricard, M.; Klocke, F.; Dambon, O.; Mullany, B. A., Journal **CIRP Annals 52 (2)**, 611-633 (2003)

[2] Kim, H.J; Kim, H.Y; Jeong, H.D; Lee, E.S; Shin, Y.J, Journal **Journal of Materials Processing Technology**, 130-131 (2002)

https://doi.org/10.1051/epjconf/201921501005

Topology optimization and additive manufacturing of an optical housing for space applications

Nils Heidler[1*], *Henrik* von Lukowicz[1,2], *Enrico* Hilpert[1,2], *Stefan* Risse[1], *Lucas* Alber[3], *Jan* Klement[3], *Frank* Heine[3], *Ralf* Bölter[3] and *Josep Maria Perdigues* Armengol[4]

[1]Fraunhofer Institute for Applied Optics and Precision Engineering IOF, 07745 Jena, Germany
[2]Institute of Applied Physics, Friedrich-Schiller-University, 07745 Jena, Germany
[3]Tesat-Spacecom GmbH & Co. KG, 71522 Backnang, Germany
[4]ESA/ESTEC, 2200 AG Noordwijk ZH, Netherlands

Abstract. The design of an optical housing for laser telecommunication in space is improved by topology optimization. Different mechanical and thermal boundary conditions are considered while minimizing the overall weight of the housing. As a proof-of-concept study, a complex and lightweight housing is made by additive manufacturing with the aluminium silicon alloy AlSi40. Post processing steps include a thermal treatment, cleaning and a mechanical machining process. Final characterization tests include the evaluation of material characteristics by tensile tests, a computed tomography scan and a CMM measurement. The final shock and vibrational test is used to proof the performance of the housing for future space applications.

1 Introduction

Opto-mechanical structures used in aerospace applications usually have to meet challenging and partially contradicting requirements. This study aims to demonstrate exemplarily that topological optimization combined with 3D printing can successfully decrease the weight of an opto-mechanical housing inside a laser communication terminal while keeping the existing requirements concerning thermal conductivity and mechanical stability.

Conventional machining processes restrict the possible design of housings, where the high freedom of design of additive manufacturing processes offer new concepts for optimized components [1,2].

The initial design of the housing is made out of AlBeMet, a beryllium and aluminium metal matrix composite material. This material offers very good mechanical properties but due to its carcinogenic properties, the production process involves difficulties and risks. As alternate material, the aluminium silicon alloy AlSi40 with 40 weight percent silicon offers good mechanical properties and the coefficient of thermal expansion is close to AlBeMet [3]. This avoids a thermo-elastic redesign of the whole opto-mechanic system.

The optimization process aims on a lightweight design with additional mechanical and thermal boundary conditions. Starting with an envelope model of the housing, different topology optimized model variants arise. Based on the overall performance, the best model variant is processed by SLM. Afterwards, thermal as well as mechanical post processing steps and metrological characterizations of the housing finalize the study.

2 Topology Optimization

The basis of the topology optimization is the envelope model of the optical housing, defining the necessary mechanical mounting interfaces for the optical components and other structural parts, see Fig.1. Within the volume of the envelope, the simulation program ANSYS is allowed to alter the material distribution to optimize the design. The applied boundary conditions involve the overall mass, geometrical, mechanical and thermal constraints. Load cases for quasi-static acceleration, sinus vibration, random vibration, shock, eigen-frequencies and static and transient thermal constraints are given. After one optimization step, all load cases are verified based on FEM simulations of the obtained design variant. The final evaluation is done by verifying if all initial conditions are met.

Fig. 1. CAD illustration of envelope model.

As exemplarily result, the first natural eigen-frequency of the final design variant occur at approx. 2850 Hz, see Fig. 2. This is well above the specified minimum frequency and above the typical incitation range for random vibration. The other given specifications are met as well.

* Corresponding author: Nils.Heidler@iof.fraunhofer.de

Fig. 2. Result of FEM simulation (modeshape of first eigen-frequency at 2850 Hz) of the optimized model.

3 Additive Manufacturing

The preparation of the optimized model for the AM process involves the addition of allowance material. This is necessary to provide enough material for the mechanical machining of the functional surfaces during the post processing steps. Using the software "Materialise Magics", the orientation of the CAD model within the building volume of the AM machine is set, supporting structure for the mechanical and thermal connection to the baseplate is defined and finally the slicing of the complete model into thin layers is accomplished [4].

For the additive manufacturing of the model, a powder-bed based, selective laser melting process is used. As AM machine, the Concept Laser M2 uses a fibre laser source with up to 400 W power and a heatable building chamber. Spherical aluminium silicon powder (AlSi40) with diameters between 20 µm and 60 µm is processed under protective gas. The used process parameters are feasible to generate volume parts with a low porosity < 0.05 % [4]. Fig. 3 shows the optimized model after the finished AM process. Additional tensile specimens are manufactured to facilitate material tests.

Fig. 3. Optical housing and tensile specimens made of AlSi40 after the AM process.

4 Post Processing and Characterization

After the AM process, the whole building plate including all parts is thermal treated to minimize the residual internal stress. An ultrasonic cleaning step is applied to remove the adhering powder particles from the surface. No further surface modification such as sandblasting or polishing are used. The optical housing is finished on a 5-axis milling machine by removing the allowance material and adding tapped holes and holes with fitting tolerances, see Fig. 4. These features cannot be realized during the AM process due to the typical surface roughness between 5 µm RMS and 20 µm RMS. Tactile measurements on a coordinate measurement machine are used to control the correct position of all machined interfaces.

A computed tomography (CT) scan is performed for quality assurance. No defects or pores are visible within the resulting voxel resolution of 120 µm.

The tensile test results reveal the good mechanical properties of the AlSi40 material with a tensile strength of approx. 260 MPa and a Young's Modulus of approx. 100 GPa. Compared to conventional AlSi40 base material the increased tensile strength of AM made AlSi40 can be explained by the finer grain structure of the material [5].

A shock and vibration test is planned to verify the FEM simulations.

5 Conclusion and Outlook

This study presents the topology optimization of an optical housing for space communication and the additive manufacturing of the resulting design out of AlSi40. All thermal and mechanical specification have been fulfilled and the overall weight was reduced by 20% compared to the initial AlBeMet housing. A CT scan, tensile tests and a shock and vibration test, verifies the good properties of the housing. Further material test shall verify the stability of the process and its application for future space missions.

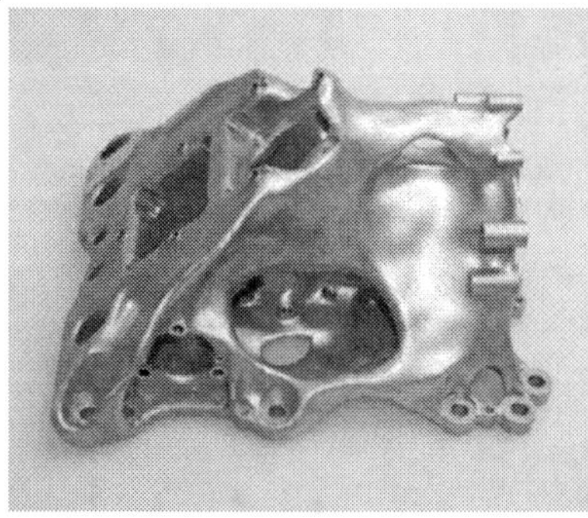

Fig. 4. Topology optimized optical housing, made by AM out of AlSi40

References:

1. Patent WO2013/017144 A1, *Method for manufacturing a mirror comprising at least one cavity and optical mirror*
2. E. Brandl, U. Heckenberger, V.Holzinger, D.Buchbinder, Mat.&Des. **34**, p. 159-169, (2012)
3. J. Kinast, K. Grabowski, A. Gebhardt, R.R. Rohloff, S. Risse, A. Tünnermann, Proc. of SPIE Vol. 10563, (2014)
4. E. Hilpert, J. Hartung, H.v. Lukowicz, T. Herffurth, N. Heidler, Optical Engineering (to be published)
5. E. Hilpert, S. Risse, MS&T (2015)

Efficient Robust Design Optimization of Optical Systems

Stephanie Kunath[1,*]

[1]Dynardo GmbH, Steubenstrasse 25, Weimar, 99423, Germany

Abstract. To accelerate the virtual product development of using optical simulation software, the Robust Design Optimization approach is very promising. Optical designs can be explored thoroughly by means of sensitivity analysis. This includes the identification of relevant input parameters and the modelling of inputs vs. outputs to understand their dependencies and interactions. Furthermore, the intelligent definition of objective functions for an efficient subsequent optimization is of high importance for multi-objective optimization tasks. To find the best trade-off between two or more merit functions, a Pareto optimization is the best choice. As a result, not only one design, but a front of best designs is obtained and the most appropriate design can be selected by the decision maker. Additionally, the best trade-off between output variation of the robustness (tolerance) and optimization targets can be found to secure the manufacturability of the optical design by several advanced approaches. The benefit of this Robust Design Optimization approach will be demonstrated.

1 Motivation

The optimization of advanced optical designs is very challenging due to their complexity, nonlinearity, a huge number of input parameters and interactions between them. At the same time, the demands for the performance of these systems are versatile and very high and even get higher, not only concerning optimization targets but also robustness criteria. Furthermore, totally new developments, like new materials, manufacturing possibilities and very short product development times, at the same time, require advanced methodologies to develop competitive optical products.

To accelerate the virtual product development using optical simulation software, additional Robust Design Optimization methods are available in Dynardos software optiSLang.

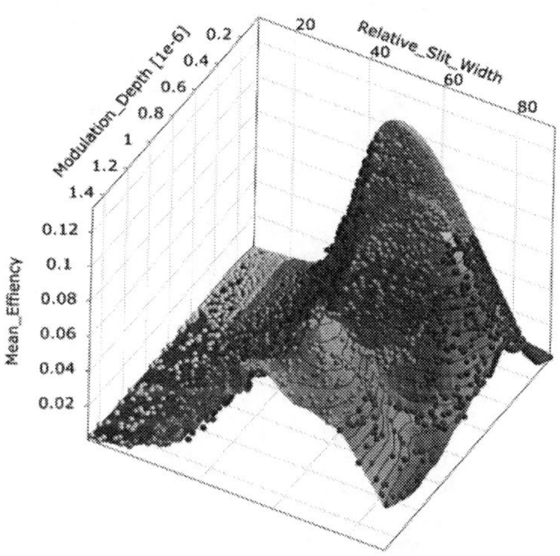

Fig. 1. Metamodel of Optimal Prognosis that shows the influence of the two input parameters Modulation Depth and Relative Slit Width on the Mean coupling Efficiency of the binary grating.

2 Robust Design Optimization Methods

2.1 Sensitivity analysis

Optical designs can be explored thoroughly by means of sensitivity analysis. This includes the identification of relevant input parameters and the (meta)modelling of inputs vs. outputs to understand their dependencies and interactions. In Fig. 1 a metamodel for the Mean Efficiency of a binary grating to couple a set of plane waves into a waveguide is shown as an example. It can be seen, that the coupling efficiency can be maximized by the Modulation Depth and Relative Slit Width at the same time. Dynardo's Metamodel of Optimal Prognosis has the ability to predict data points that haven't been used for modelling [1,2].

2.2 Optimization

The intelligent definition of objective functions for an efficient subsequent optimization is of high importance for multi-objective optimization tasks. Questions that can be answered using the results of the sensitivity analysis are:

- How to weight the objectives that are not in conflict to define them in one objective function? This can be defined based on the variation of the merits as result of the sensitivity analysis.

* Corresponding author: stephanie.kunath@dynardo.de

- Are the objectives (merits) in conflict? In this case, they should be defined in separate objective functions.

To find the best trade-off between two or more merit functions, a Pareto optimization is the best choice. As a result, not only one design, but a front of best designs is obtained and the most appropriate design(s) can be selected by the decision maker. These best designs are usually plotted as Pareto front as shown in Fig. 2. Additionally, further understanding of the optical design can be obtained by clustering interesting parameters and colouring each of these resulting clusters. In Fig. 2 a cluster analysis of the Fill Factor of a slanted grating was considered to understand its influence on the performance (Uniformity Error as well as Mean Efficiency to couple a set of plane waves into a waveguide). It is obvious, that high fill factors lead to lower efficiencies. Furthermore, the trade-off between Uniformity Error and Mean Efficiency gets visible: the higher the efficiency the higher the uniformity error.

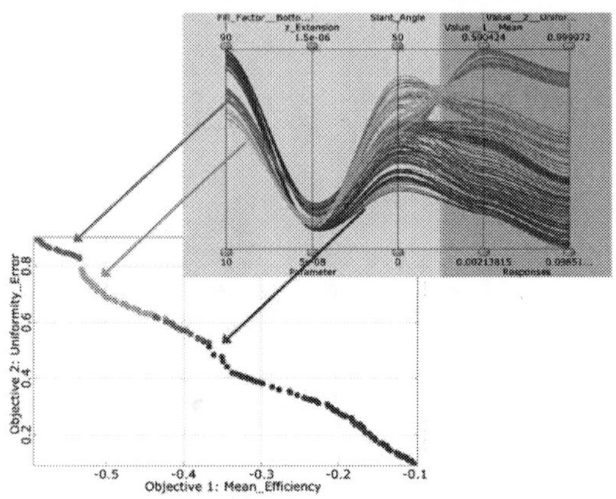

Fig. 2. Anthill plot with Pareto front designs (lower left plot) representing the trade-off between Uniformity Error and Mean Efficiency of a slanted grating. The colours correspond to the colour of the cluster analysis illustrated in the parallel coordinates plot (upper right plot). Here, the Fill Factor was separated into 3 clusters.

2.3 Robustness analysis

Besides the optimization of an optical design, its robustness is of high importance. That means, the optical performance should be rather independent on input parameter scattering due to tolerances, material parameter variation, heating etc. Therefore, the best trade-off between optimization and robustness targets needs to be found to secure the manufacturability of the optical design. Several strategies for Robust Design Optimization are available:

1. Iterative approach: first optimization, and then robustness analysis to
   - reduce critical input scattering and/ or
   - perform a subsequent optimization with changed constraints to find another, more robust optimum
2. Coupled approach: the optimizer contains optimization <u>and</u> robustness criteria as
   - constraint or
   - second objective function (Pareto optimization).

The results of such a coupled and fully automated Robust Design Optimization approach using a second objective function is demonstrated in Fig. 3. Here, a Cooke triplet was used as example. With this approach, the optical designer can make the final decision about the best optical design with other decision makers together by analysing the trade-off between optimization and robustness criteria. In this example, it gets obvious that the higher the performance in terms of optimization criteria, the worse the robustness of the design (higher values indicate higher variation of the performance).

With this approach, many iterative design steps can be saved. As a consequence, the multidisciplinary work between engineers of different physical domains, managers and the customer will be more efficient and transparent.

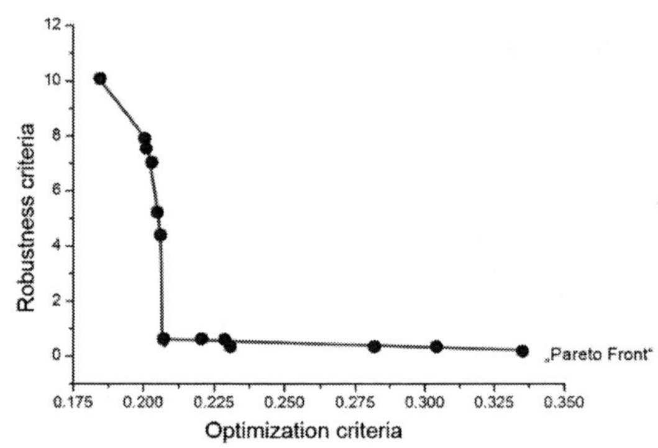

Fig. 3. Pareto Front as a result of a coupled Robust Design Optimization of a Cooke triplet. Two objective functions were used: one for the robustness criteria (here: weighted standard deviations of the merits) and a second one for the optimization criteria (here: weighted merits). Each design point on the front is a best design as a compromise between the two objectives.

References

1. Most, T. and J. Will (2011). Sensitivity analysis using the Metamodel of Optimal Prognosis, Proceedings Weimarer Optimization and Stochastic Days 8.0, Weimar, Germany
2. Most, T. and J. Will (2012). Robust Design Optimization in industrial virtual product development, Proceedings 5th International Conference on Reliable Engineering Computing, Brno, 2012

EPJ Web of Conferences **215**, 02002 (2019)
EOS Optical Technologies
https://doi.org/10.1051/epjconf/201921502002

Multifunctional contamination-resistant coatings

Nadja Felde[1,*], *Anne* Gärtner[1], *Stefan* Schwinde[1] and *Sven* Schröder[1]

[1]Fraunhofer Institute for Applied Optics and Precision Engineering, Albert-Einstein-Strasse 7, 07745 Jena, Germany

Abstract. Surface contaminations can critically affect the performance of optical surfaces, in particular with respect to light scattering, optical losses, and laser stability. Thus, avoiding contaminations and reducing contamination-induced effects is of particular interest in the manufacturing of optical systems. By combining a specific thin film design with a tailored structural design, contamination-resistant coatings with a high optical quality can be realized. Most important is the balance of self-cleaning- and light scattering-relevant surface roughness components.

Surface contaminations of optical components can critically affect the performance of the entire optical system. Typical contaminations, which are most relevant in optical applications, can be for example dust particles from the environment or fluids from the manufacturing process. However, over the last years also bio-contaminations have become increasingly critical.

To avoid contamination-induced effects, such as increased optical losses or reduced laser stability, optical coatings have to be applied, which reduce the accumulation of contaminations as well as their optical effects while keeping the optical quality at a high level.

Addressing this issue, a multifunctional coating was developed, which is characterized by the following functionalities (Fig. 1):

- Self-cleaning/contamination-repellent properties by wetting-relevant structuring
- Contamination-insensitivity by tailored optical coating design
- Broadband antireflective characteristics by structures with graded refractive index
- Low scattering by shifting the function-relevant structures to lateral scales, which are less relevant for light scattering.

For the self-cleaning/contamination repellent behaviour, the minimization of the adhesion forces between the surface and the contaminating particle or fluid is of particular importance. This is realized by tailoring the wetting properties of the coating to hydrophobic and oleophobic, severely water- and oil-repelling, properties. For this, a defined structuring and chemical functionalization [1] of the coating surface is needed. However, although required for the functionality, the increased surface roughness may have also a negative impact on the optical performance. This is the case as surface roughness leads to increased light scattering [2,3].

To address this problem, topographic, functional, and light scattering characterization methods have to be combined to individually analyse the specific properties and to comprehensively clarify their correlations to each other. This information is needed to selectively define the wetting- and light scattering-relevant spatial frequency range, which is based on power spectral density functions. This allows appropriate surface roughness structures of the multifunctional coating to be designed, which includes the shifting of the wetting-relevant surface structures to lateral scales that reduce or even prevent scatter losses [1, 4].

Nonetheless, complete avoiding of contaminations is inevitable over time. Thus, the multifunctional coating can further be equipped by an additional specific optical design. Within this design, adopting the layer thickness, the standing electric field strength at the position of the contaminating particle or fluid is modified [6,7]. Thus, contamination-induced light scattering properties can effectively be tailored.

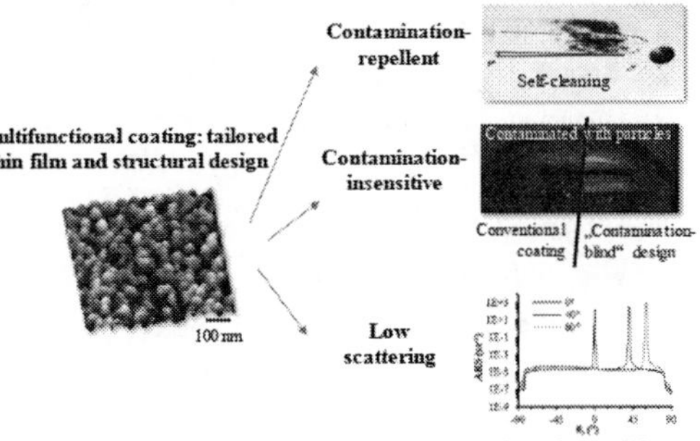

Fig. 1. Multifunctional contamination-resistant coating.

---
* Corresponding author: nadja.felde@iof.fraunhofer.de

The resulting multifunctional coating can significantly enhance the wetting properties of a smooth optical surface. Compared to a smooth surface with functionalization, reaching only advancing and receding contact angles of 110° and 95°, the multifunctional coating exhibits advancing and receding water contact angles of 155° and 145°. Thus, water drops roll off at a tilt angle of the surface of 3°, ensuring particle contaminations to be removed [1].

Such repellence holds also true for contaminations with oil-containing liquids [8,9] such as human fingerprints and can even be transferred to reduce contaminations with molten metal debris [10]. However, with respect to bio-contaminations both states, bacteria-adhesion and bacteria-colonization, have been shown to be impacted.

The main point, however, is that the optical quality of the components can be preserved after coating. In case of the multifunctional coating designed for applications in the visible spectral range the scatter losses are as low as 90 ppm in the reflectance direction and results in less than 75 ppm in the transmittance direction. Due to the function-relevant structuring additional broadband antireflective properties can be achieved.

References

1. N. Felde, L. Coriand, S. Schröder, A. Duparré, A. Tünnermann, Adv. Opt. Technol. **6**, 387–394 (2017)
2. S. Schröder, M. Trost, M. Garrick, A. Duparré, X. Cheng, J. Zhang, and Z. Wang, Thin Solid Films **592**, 248–255 (2015)
3. J. C. Stover, *Optical Scattering: Measurement and Analysis* (SPIE Press, Bellingham, Wash., 2012)
4. N. Felde, L. Coriand, A. Duparré, A. Tünnermann, J. Coat. Sci. Technol. **3**, 100–108 (2016)
5. Patent application DE102014112133 A1, 2014
6. A. von Finck, S. Wilbrandt, O. Stenzel, S. Schröder, Adv. Opt. Technol. **6**, 371–378 (2017)
7. Patent application DE 102016103339 A1, 2016
8. T. Oberbillig, N. Felde, M. Franz, L. Coriand, N. Dahms, K. Munderloh, K. Schultz, M. Fliedner, A. Duparré, Opt. Photonik **13**, 14–18, (2018)
9. L. Coriand, N. Felde, and A. Duparré, *"Wettability Behavior of Oleophilic and Oleophobic Nanorough Surfaces in Air or Immersed in Water," in Advances in Contact Angle, Wettability and Adhesion*, (John Wiley & Sons, Inc., Hoboken, New York, 2018)
10. Patent application DE102018110251, 2018.

 https://doi.org/10.1051/epjconf/201921502003

Multilevel axicon for perfect optical vortex generation

Rebeca Tudor*, *Mihai* Kusko, *Cristian* Kusko, and *Andrei* Avram

National Institute for R&D in Microtechnology IMT, Bucharest, 077190, Romania

Abstract. We present the fabrication of a beam shaper with 32 levels for the generation of nondiffractive optical fields representing quasi-Bessel beams of order zero. This optical element is designed for visible light (λ=633 nm) and fabricated using standard photolithography and a fine calibrated reactive ion etching process. A large number of levels approximates a continuous conical surface so that the optical quality of the element is very good. It is investigated the possibility of generating perfect optical vortices with this class of optical elements.

1 Introduction

Ideal Bessel beams are nondiffractive optical fields described by invariant intensity profiles along an infinite propagation distance. In experiments, one can generate only quasi-Bessel beams characterized by finite focal extensions and intensity profiles that vary along the propagation axis but only after a specific distance called focal length. Quasi-Bessel beams can be generated with conical lens, planar Pancharatnam-Berry phase elements, spatial light modulator, Fresnel lens-based elements [1-4]. Nondiffractive unique beams are widely exploited in fields like optical trapping [5], beam shaping [6-9] optical communications [10-11], microscopy [12].

2 Fabrication process

In this work we designed, fabricated and characterized a diffractive axicon as a phase control element in order to obtain nondiffractive optical fields for visible light. The design was considered an approximation of continuous conical surface in order to generate a Fresnel based element presenting a discrete surface defined by 32 levels. This element was fabricated on a silicon wafer using standard photolithography and a fine calibrated reactive ion etching process. The multilevel diffractive optical element generates quasi-Bessel beams of order zero.

2.1. Structural characterization

We investigated the optical quality of the fabricated axicon with the optical microscope. In Fig. 1 is represented a small area of an axicon of order zero with 32 levels. Red colour represents the maximum level and the green colour indicate the minimum level. The step height between the maximum and minimum level introduces a phase shift in the wavefront of an incident Gaussian beam in order to generate nondiffractive beams.

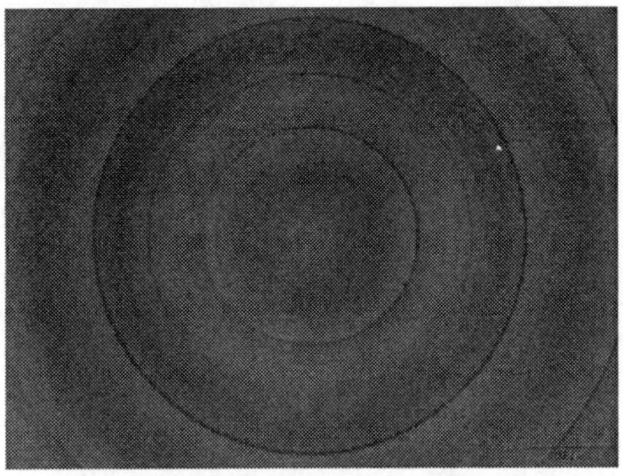

Fig. 1. Optical image of an axicon of order zero with 32 levels

2.2 Functional characterization

We beam shaped a Gaussian beam with the wavelength λ=633 nm using the fabricated axicon and generated a quasi-Bessel beam with order zero which is illustrated in the Fig. 2. One can observe the intensity distribution characterized by a central lobe surrounded by concentric rings.

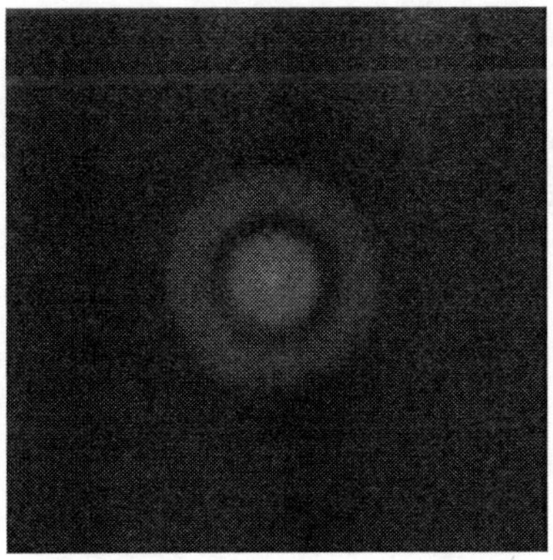

Fig. 2. Intensity distribution for a Bessel beam of order zero

* Corresponding author: rebeca.tudor@imt.ro

2.3 Conclusions

We designed, fabricated and characterized a multilevel beam shaper for visible light (λ=633 nm) in order to generate a quasi-Bessel beam of order zero. We obtained a diffractive axicon with a very good approximation of a continuous surface by increasing the number of levels. This was possible due to multiple photolitographical and reactive ion etching processes. The structural and functional characterization confirmed the optical quality of the fabricated axicon with 32 levels. This optical element has potential industrial application in the generation of perfect optical vortices [10], beam shaping, optical communications and optical imaging.

References

1. I. Golub, Opt Lett. **31**(12), (2006)
2. Y. Liu, Y. Ke, J. Zhou, Y. Liu, H. Luo, S. Wen, D. Fan, Sci. Rep **7** (2017)
3. R. Li, X. Yu, T. Peng, Y. Yang, B. Yao, C. Zhang, T. Ye, J. Opt. **20** 085603 (2018)
4. D.J. Fischer, C.J. Harkrider, D.T. Moore, Appl. Opt., AO. **39** 2687–2694 (2000)
5. V. Garces-Chavez, D. McGloin, H. Melville, W. Sibbett, K. Dholakia Nature **419** (2002)
6. L. Shi, X. Dong, Q. Deng, Y. Lu, Y. Ye, C. Du, Opt. Eng **50** (2011)
7. D. Deng, Y. Li, Y. Han, X. Su, J. Ye, J. Gao, Q. Sun, S. Qu, OE. **24** (2016)
8. S. Khonina, S. Degtyarev, D. Savelyev, A. Ustinov, OE. **25** (2017)
9. S. Supp, J. Jahns, JEOS-RP. **14** (2018)
10. A. Burvall, P. Martinsson, A.T. Friberg, OE. **12** (2004)
11. S. Chen, S. Li, Y. Zhao, J. Liu, L. Zhu, A. Wang, J. Du, L. Shen, J. Wang, OL. **41** (2016)
12. F. O. Fahrbach, A. Rohrbach Opt. Exp **18** (2010)

Acknowlegdement

The authors would like to acknowledge the financial support from IMT-Bucharest Core Programme "MICRO-NANO-SIS" PN 4N/2018 and "MICRO-NANO-SIS-PLUS" PN 14N/2018 supported by MCI.

Towards accessible nanophotonics: multimode interferometer on strip-loaded slot waveguide

M. Roussey[1*], *R. Rao*[1], *and S. Pélisset*[1]

[1]University of Eastern Finland, Institute of Photonics, Joensuu, FI-80100, Finland

Abstract. Nano- and micro-photonics are the key-enabling tools for future integrated components and circuitry operating at low power and high speed. By using a strip-loaded platform, we show how we can dramatically reduce the complexity, in terms of fabrication and tolerances, of the most advanced devices. Different configurations of multimode interference devices are presented. We show the design, fabrication, and optical characterization of these components.

1 Introduction

A strip-loaded waveguide consists of a high refractive index thin film, in which light is confined laterally by a strip patterned on to of it [1]. This structure offers a high level of degree of freedom in terms of materials, features, and fabrication constraints. The guiding layer can therefore be of any kind of material since it is not patterned and does not need to be processed after deposition. Lithium niobate, glasses, titanium dioxide strip loaded waveguides have been demonstrated successfully in the past years. The loading strip must have a refractive index lower than the effective index of the guided slab mode in order to avoid leaking of light in it. Polymer strips are ideal due to their relatively low refractive index and easiness to pattern. We have recently demonstrated that such a strip-loaded waveguide can be easily combined with a horizontal slot waveguide [2] in order to confine at the nanoscale the electric field in a low refractive index layer sandwiched between two high refractive index films [3,4].

A strip-loaded waveguide is a low refractive index contrast waveguide. This implies larger features, i.e., easier fabrication, and a higher tolerance to imperfections due to fabrication.

2 Results and discussion

In this work, we present a multimode interference (MMI) device on a strip-loaded horizontal slot waveguide (SLSW) platform. The waveguide is composed of two layers of titanium dioxide (nTiO2 = 2.27 at λ = 1550 nm) of thickness 200 nm separated by an 80 nm layer of silica (nSiO2 = 1.444). The loading strip is made of AZ nLOF, an electron beam resist. Figure 1a) depicts, in a top view, the geometrical parameters of the structure. The waveguides width is set to 1.2 µm to achieve single mode condition. However, to smooth the coupling to the free propagation region, the waveguides are tapered to 5 µm-width. Several devices have been realized (1x2, 1x3, 1x4, and 2x2). Only the 1x2 is presented in Figure 1.

Finite Different Time Domain (FDTD) simulations have been used to calculate both the modes and the behavior of the field inside the devices. Because of the low index contrast of the waveguide, we were able to use the effective index approximation for the quasi-TM mode (fundamental slot mode) allowing a more refined mesh in a two-dimensional simulation. The field intensity is presented in Figure 1b). We can clearly see the interference pattern in the free propagation region leading to the splitting at the output.

The devices were fabricated by atomic layer deposition for the slot waveguide and electron beam lithography for the patterning of the strip. Figure 1c) shows images obtained with a microscope at x20 and x50 magnifications. We can remark that some stitching between the waveguides and the free propagation region appears. This is due to the fact that two different doses have been used for the patterning in order to obtain the best quality in terms of corner of the MMI and fastest writing speed. Using tapers, allows reducing the effect of such a defect in the patterning.

The devices were characterized using tapered lens fibers to inject light and a x20 microscope objective and an infrared camera to observe the output modes. Figure 1d) shows this result proving the correct operation of the device despite the defects occurring during the fabrication.

* Corresponding author: matthieu.roussey@uef.fi

https://doi.org/10.1051/epjconf/201921502004

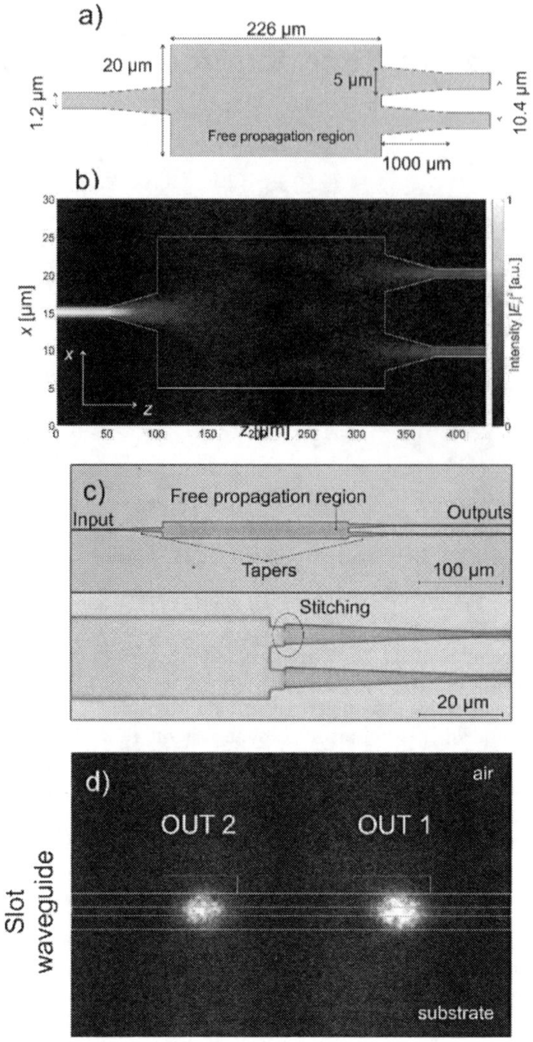

Fig. 1. a) Sketch and geometrical parameters of a 1x2 MMI device on SLSW platform. b) FDTD simulation of the device. The intensity distribution viewed from top is displayed. c) Micrographs of the fabricated device (top: 20x, bottom: 50x). d) Picture taken from the output of the device when illuminated under TM polarization at $\lambda = 1550$ nm.

3 Conclusions

Using the SLSW platform to create an MMI device proves that we can produce standard key building blocks of integrated photonics using reliable fabrication techniques. It is to be noted that such devices can also be fabricated by mass-production methods instead of electron beam lithography, such as nano-imprinting as we demonstrated in [5].

The work is part of the Academy of Finland Flagship Programme, Photonics Research and Innovation (PREIN), decision 320166.

References

1. H. Furuta, H. Noda, A. Ihaya, "Novel optical waveguide for integrated optics," Appl. Opt. **13**, 322–326 (1974).
2. V.R. Almeida, Q. Xu, C.A. Barrios, M. Lipson, "Guiding and confining light in void nanostructure," Opt. Lett. 29, 1209–1211 (2004).
3. M. Roussey, L. Ahmadi, S. Pélisset, M. Häyrinen, A. Bera, V. Kontturi, J. Laukkanen, I. Vartiainen, S. Honkanen, M. Kuittinen, "Strip-loaded horizontal slot waveguide," Opt. Lett. **42**, 211–214 (2017).
4. S. Pélisset, J. Laukkanen, M. Kuittinen, S. Honkanen, M. Roussey, "Modal properties of a strip-loaded horizontal slot waveguide," J. Eur. Opt. Soc. – Rapid Publ. **13**:37 (2017).
5. L. Ahmadi, V. Kontturi, J. Laukkanen, S. Honkanen, M. Kuittinen, M. Roussey, "Strip-loaded waveguide on titanium dioxide thin films by nanoimprint replication", Optics Letters, **42**(3) 527-530 (2017).

https://doi.org/10.1051/epjconf/201921503001

Direct Laser Fabrication of Cylindrical Lenses in Wide Band Gap Materials

Sebastian Buettner, Michael Pfeifer,* and *Steffen Weissmantel*

Laserinstitut Hochschule Mittweida, 09648, Mittweida, Germany

Abstract. The Fluorine laser microstructuring technique enables the direct laser fabrication of cylindrical lenses and lens arrays thereof in wide band gap materials. For the mask projection technique, a special mask geometry was calculated, which allows the fabrication of cylindrical lenses with a nearly optimum curved surface. Based on our results of processing fused silica and borosilicate glass, we investigated the possibility to apply these microstructuring technique to wide band gap materials like calcium fluoride. The radius of curvature (ROC) can be adjusted by the process parameters laser pulse fluence and pulse-to-pulse overlap in a range of 130 to 450 μm micrometer. A minimal surface roughness of 100 nm RMS can be reached.

1 Introduction

Micro optics are becoming more and more important to deal with technical problems. A wide range of techniques were developed to manufacture optical elements in the micrometer range in the last few years and in particular for the fabrication of cylindrical lenses and lens arrays [1, 2]. One of these is the fluorine laser microstructuring. This technique enables the direct laser fabrication of micro lenses in glass and especially in wide band gap materials like calcium fluoride. This is achieved by the short wavelength of 157 nm and the corresponding high photon energy of 7.9 eV of the fluorine laser. The microstructuring is done by a fully automated high-precision laser micromachining station. The pulse duration of the laser is 25 ns, the maximum repetition rate is 200 Hz and the maximum pulse energy is 30 mJ.

Based on our research on the microstructuring of blaze gratings [3], diffractive optical elements (DOEs) [4], and Fresnel lenses [5] we further developed these technique for the fabrication of cylindrical lenses with a nearly optimal shaped lens surface and a defined ROC. In a first step we fabricated cylindrical lenses and arrays thereof in fused silica and borosilicate glass [6]. In combination with the CO_2 laser smoothing as a follow up treatment process, we can fabricate these kind of optical elements with a surface roughness in a range of a few nanometers. In a next step we applied this technique to wide band gap materials.

2 Results and Discussion

Because of the short wavelength of the fluorine laser the microstructuring of wide band gap materials is possible. However, even materials with a higher band gap than the photon energy can be processed, as can be seen in Fig. 1. The fabrication of cylindrical lenses requires the knowledge of the specific material parameters ablation threshold H_{th} and absorption coefficient α. These parameters can be calculated using the Lambert-Beer's law and the ablation depth per pulse depending on the laser pulse fluence. Therefore, a parameter study for every used material has been done. In Fig. 1 the ablation depth per pulse depending on the laser pulse fluence is shown for different wide band gap materials.

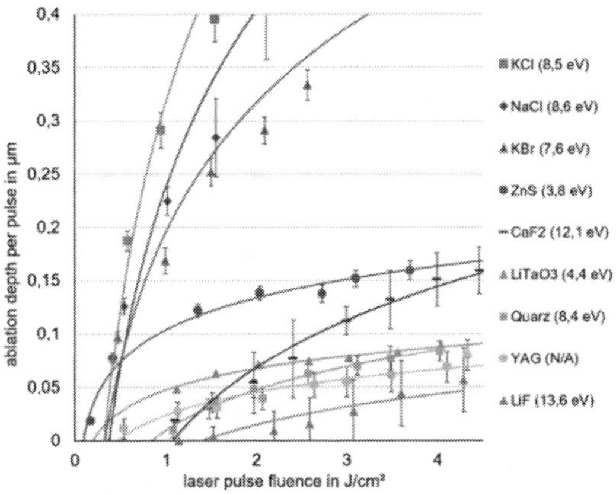

Fig. 1. Ablation depth per pulse for different wide band gap materials.

For the demonstration of the fabrication of cylindrical lenses in wide band gap materials, we choose calcium fluoride as material for our investigations. The determined ablation threshold of this material is $H_{th} = 1.07$ J/cm² and the absorption coefficient is $\alpha = 7.861 \cdot 10^4$ cm^{-1}. Furthermore, we used the mask, which we calculated for previous investigations on fused silica and borosilicate glass. A micrograph of this mask is shown in Fig. 2.

* Corresponding author: buettner@hs-mittweida.de

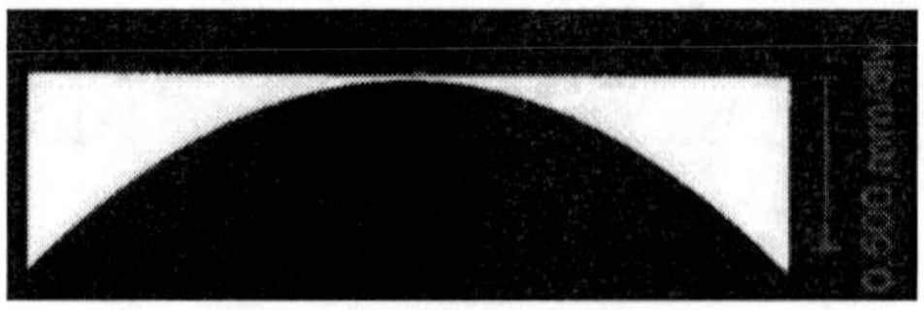

Fig. 2. Optical micrograph of the calculated mask.

For the fabrication of cylindrical lenses in calcium fluoride, the laser pulse fluence was varied in a range of 2 to 4.5 J/cm² and the pulse-to-pulse overlap in a range of 95 to 99 %. In general, the ROC of the lenses depending on the laser pulse fluence and the pulse-to-pulse overlap. The pulse-to-pulse overlap is defined by the relation of the maximum mask aperture to the pulse-to-pulse distance. For this material, the fabrication of nearly optimal shaped lenses requires a defined laser pulse fluence of 3.5 J/cm². At this laser pulse fluence, the course of the profile section is in a good agreement with the best fit ROC, as can be seen in Fig. 3. The ROC of these lenses are within the range of 130 to 450 µm. Furthermore, the surface roughness of these lenses is in a range of 100 nm RMS.

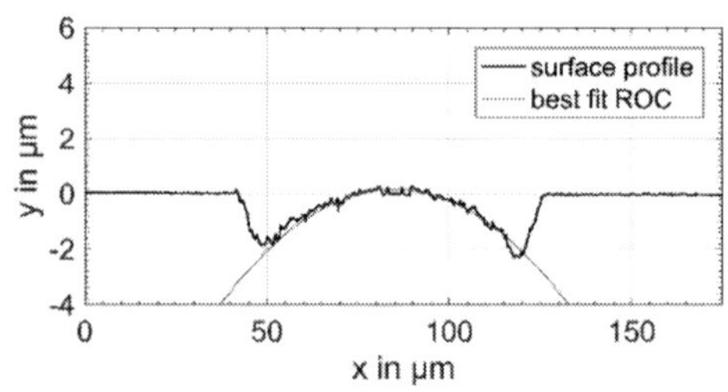

Fig. 3. Profile section of a cylindrical lens fabricated in calcium fluoride (H = 3.5 J/cm², overlap 98 %).

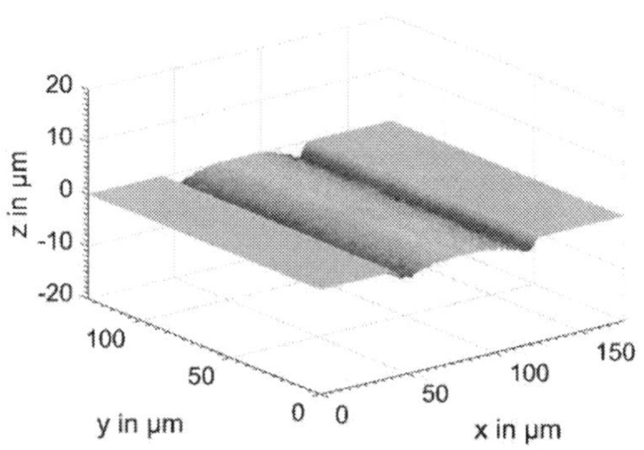

Fig. 4. Three dimensional micrograph of a cylindrical lens fabricated in calcium fluoride (H = 3.5 J/cm², overlap 98 %).

For other laser pulse fluences, the profile section deviates from the circular form. This is caused by a strong increase of the surface roughness at higher and an inconvenient ablation behavior at lower laser pulse fluences. The reason for the latter is the high ablation threshold, which lead to an incomplete removal of the material at the vertex of the lenses. This problem can be solved by adapting the mask geometry in the region of the vertex. However, the wave-optical simulation show a focussing effect of the lenses. For this, the measured three dimensional surface data were used to calculate the phase shift and the electrical field strength of a propagating electromagnetic wave. Due to the roughness of the surface, a part of the light is scattered, as can be seen in Fig 5.

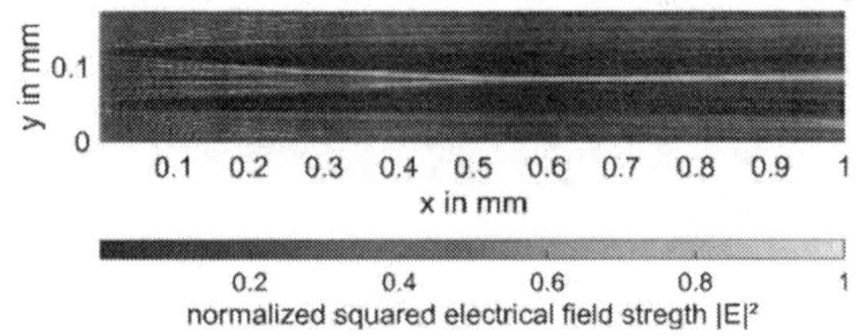

Fig. 5. Squared electrical field strength $|E|^2$ normalized to 1 depending on the propagation direction x, calculated from a profile section of a cylindrical lens fabricated in calcium fluoride (λ = 532 nm, n = 1.43).

3 Conclusion

The fabrication of cylindrical lenses in calcium fluoride can be done using the fluorine laser microstructuring technique. At a laser pulse fluence of 3.5 J/cm², lenses with a nearly optimal curved surface can be fabricated, with an ROC in the range from 130 up to 450 µm. In general, the ROC depending on the pulse-to-pulse overlap. For lower and higher laser pulse fluences, the surface profile section deviates from the circular form. In particular at lower fluences, the microstructuring process is affected by the incomplete removal of the material. Regarding to this, the calculation of an adapted mask geometry can enlarge the field of applicable parameters. This may enables the fabrication of lenses at a lower laser pulse fluence, which can lead to a reduction of the surface roughness. Furthermore, we want to use an ion beam for follow-up treatment. In this way, we can reach a further improvement of the lens surface.

Acknowledgements

The authors gratefully acknowledge the financial support of the present work by the European Union and the Free State of Saxony.

Europäische Union
Europa fördert Sachsen.
ESF
Europäischer Sozialfonds

References

[1] Z. Luo, K. Yin, X. Dong, J. Duan, Opt. Mat. **78** (2018)
[2] H.-K. Choi, M.S. Ahsan, D. Yoo, I.-B. Sohn, Y.-C. Noh, J.-T. Kim, H.-M. Kang, Opt. Laser Technol. **75** (2015)
[3] M. Pfeifer, S. Weissmantel, G. Reisse, Appl. Phys. A **112**, 1 (2013)
[4] M. Pfeifer, F. Jahn, A. Kratsch, B. Steiger, S. Weissmantel, Proc. of PHOTOPTICS (2014)
[5] M. Pfeifer, S. Büttner, R. Zhang, M. Serbay, S. Weißmantel, In Sci. Rep. of Univ. of Appl. Sci. Mittweida, 2 (2017)
[6] S. Buettner, M. Pfeifer, S. Weissmantel, Proc. of PHOTOPTICS (2019)

NOPT – New polishing techniques for scalable, light-weighted mirrors of different materials

Sebastiano Spinelli[1] , *Luigina* Arcangeli[1], *Ruben* Mazzoleni[1], *Massimiliano* Rossi[1*], *Marco* Terraneo[1], Roman Windpassinger [2]

[1]Media Lario Technologies, 23842 Bosisio Parini, Italy

[2]ESA , 23842 Bosisio Parini, Italy

Abstract. In the framework of an assessment of optical polishing techniques, ESA has signed a contract with Media Lario to deliver two 250 mm mirrors with a common optical design to be polished down to very tight surface requirements. NOPT mirrors are respectively made of Zerodur and AlSi alloy with electroless nickel and will they be polished by means of bonnet polishing. Mirrors are light-weighted up to 20kg/m^2; the mechanical and optical design is proven to be scalable up to 1m CA surfaces. This paper reviews the mirrors opto-mechanical design, introduce the polishing and metrology strategy while highlighting the differences and the common point in fabricating such mirror in Zerodur and metal.

1. New Optical Polishing Techniques project

Modern polishing techniques for high performance optics often use of small area tools, in contrast to classical large area polishing techniques, where the complete optical surface to be polished or large part of is covered by the polishing tool, limiting the available surface shapes to plane or spherical ones. Using local area polishing techniques enables the polishing of aspherical and free-form surface shapes. In this framework ESA has signed a contract with Media Lario to deliver two 230 mm clear aperture (CA) mirrors polished down to very tight specifications (shown in Table 1) by means of Zeeko bonnet polishing process [1].

Table 1: NOPT mirror PSD requirements after polishing

Requirement	Integrated PSD value (nm RMS)
Shape (Low Spatial Frequency f<0.021 cy/mm)	7.5
Waviness (Mid Spatial Frequency 0.021cy/mm<f<1.1 cy/mm)	3
Roughness (High Spatial Frequency f>1.1 cy/mm)	0.7

Mirrors will be light-weighted up to an areal density of 20 kg/m^2 and will be fabricated one in Zerodur and one in AlSi aluminium alloy coated with a polishing layer of electroless Nickel. The use of the AlSi alloy, that is CTE matched with the Nickel coating, allows the fabrication of an almost athermal design of the mirror as for the case of Zerodur. The polishing process, the light weighting and mechanical design of the mirror are proven to be scalable up 1 m CA [2].

2. NOPT mirrors design

Mirrors share the optical design adopted for the primary mirror of the ESA mission CHEOPS [3], apart the reduction of the CA from 330mm to 230 mm. This design was adopted in order to build a demonstrator known as being a suitable surface to be selected in space systems. The optical surface is thus hyperbolic, with a curvature radius of 742.756mm and a conic constant k=-10.2456.

While sharing the CHEOPS optical design, the mechanical design of the mirrors is completely different from the original one.

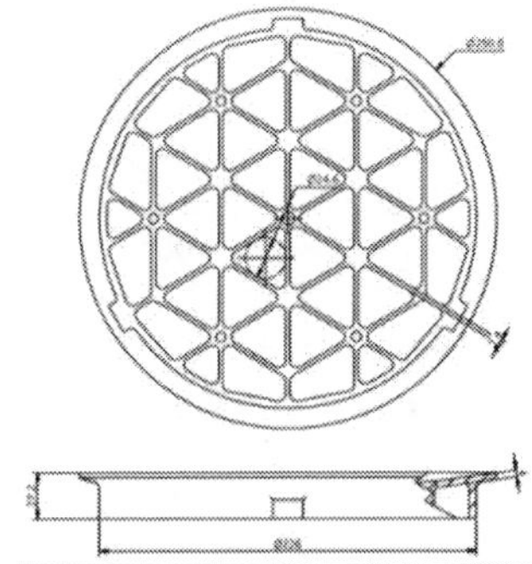

Figure 1: Skecth of the NOPT mirrors.

* Corresponding author: massimiliano.rossi@medialario.com

https://doi.org/10.1051/epjconf/201921503002

NOPT mirrors are designed to have an areal density of 20 kg/m^2 and yet a first eigenfrequency higher than 500 Hz. This has been achieved by designing a triangular shape modular pattern for the light-weighting scheme according to [4]. This approach fulfills also the requirement imposed by the ESA SoW of having a light-weighting scheme scalable up to 1m CA.

3. NOPT mirrors polishing

Mirror polishing will be carried out by means of bonnet polishing machines. Media Lario has developed through the years different polishing processes specific both for Zerodur and AlSi electroless nickel plated mirrors able to bring down the surface errors to the requirements.

As a general description, the polishing process can be described as follows

I. Sub-surface damage removal: this is a polishing phase devoted to glass parts only in which material is uniformly removed from the surface in order to avoid superficial cracks induced by the grinding process [5]. Metal mirrors do not require this phase as they are diamond turned.

II. Low frequency errors (shape) are deterministically removed from the surface. In this polishing phase the bonnet transfer function is assumed to be point-like with respect to the shape error to be removed and is effective on LSF errors of Table 1

III. Mid-spatial frequency errors are deterministically removed from the surface by means of bonnet Precession polishing [6]: in this phase, the real bonnet transfer function is measured and deconvolved from the error map in order to obtain a more accurate dwell time map on the surface from the polishing tool to attack MSF of Table 1

IV. High frequency errors are stochastically removed lapping the surface with a proper bonnet-based process, in order to match the roughness HSF specification in Table 1.

As of now, the fabrication of both mirror substrates is completed. The Zerodur mirror polishing has just started, while the metal mirror diamond turning will be completed by the end of March. The project plan is to deliver both mirrors polished down to specifications before June 2019.

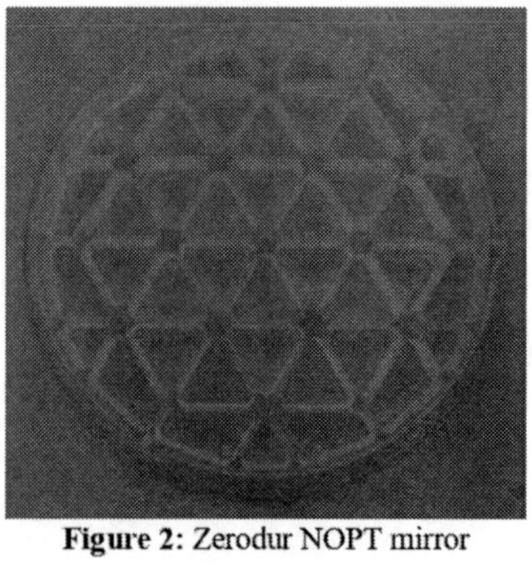

Figure 2: Zerodur NOPT mirror

4. NOPT mirrors Metrology

Media Lario has developed metrology tools to characterize the surface errors in the spatial frequency ranges indicated by mirror specifications.

Specifically, LSF and MSF will be addressed starting out from interferometric measurements with respect of a CGH. A dedicated setup for having these measurements under the available 6" WYKO interferometer that minimizes the distortion induced by the mounting was designed with FEA and fabricated.

These measurements will provide the data for PSD definition up to 1.1 cycles/mm frequency.
HSF will be measured by means of a Taylor Hobson CCI white light interferometer that has the capability to reconstruct the 2D surface PSD from 6 mm down to 1 μm spatial wavelengths.

In order to discriminate the impact of the WLI optics in the PSD reconstruction, a correlation analysis between the WLI PSD and ARS measurements will be carried out on two 4" samples made of the same materials as NOPT mirrors. ARS was measured by means of the ALBATROSS apparatus, developed by Fraunhofer IOF Jena.

References

1. R. G. Bingham et al, "A novel automated process for aspheric surfaces," Proc. SPIE **4093**, 445-448, 2000
2. ESA ITT AO/1-8769/16/NL/RA, New Optical Polishing Techniques, Statement of Work, issue 1, 14/07/2016.
3. The CHEOPS team, "CHEOPS definition study report (Red Book)," **ESA/SRE(2013)**, 2013.
4. D. Vukobratovich "Lightweight Mirror Design" in Optomechanical Engineering Handbook, CRC Press, 1999.
5. Lambropoulos, J. C., "From abrasive size to subsurface damage in grinding," OSA Technical Digest **8**, 17-18, 2000
6. Walker et al., The 'Precessions' tooling for polishing and figuring flat, spherical and aspheric surfaces, Opt. Express **11**, 958-964, 2003.

Optical freeform generation by laser machining and plasma-assisted polishing

Thomas Arnold[1,2*], *Anne* Maiwald[2], *Georg* Böhm[2], *Martin* Erhrhard[2], and *Klaus* Zimmer[2]

[1]Institut für Fertigungstechnik, Fakultät Maschinenwesen, Technische Universität Dresden, Germany
[2]Leibniz-Institut für Oberflächenmodifizierung e.V., Permoserstr. 15, 04318 Leipzig, Germany

Abstract. Tailored optical freeform lenses are required for different applications. Sub-aperture deterministic machining techniques such as plasma jet machining have shown great potential to generate freeform surfaces. However, depending on the required local slopes of the surface shape geometrical limitations occur due to the lateral tool function width. In the paper an alternative approach to fabricate freeform shapes exhibiting steep local slopes is presented. A first step involves a dwell time based fs-laser ablation process to generate the surface contour on a fused silica sample. Since the resulting roughness after laser machining lies in the range of 400 nm RMS which does not match optical requirements a subsequent plasma jet based polishing step is performed where micro-roughness is drastically reduced to values below 0.3 nm RMS.

1 Introduction

Today, an increasing demand for complex shaped optical elements like non-standard aspheres, acylinders, or freeform elements is observed. Recent advances in optical manufacturing technologies give more and more freedom to optical system designers to implement freeform surfaces. The use of freeform elements in optical systems provides new functionalities in illumination and imaging and yields less optical surfaces and thus more compact and lightweight system designs. In cases, where highly individualized and tailored single piece optics or small batches are required, replication techniques like moulding are hardly appropriate for production. Classical manufacturing of such optics generally involves intense sub-aperture polishing after a freeform grinding process to remove sub-surface damage (SSD). Depending on the grinding process material removal of 10-15 µm is necessary to get sufficient optical surface quality in the high spatial frequency range. Furthermore, sub-aperture polishing often leads to significant shape alteration, particularly in the case of significant variations of local surface curvature, which originates from the corresponding local variation of polishing removal rates. After the polishing process, surface roughness is usually within the specified range, though an unintended shape deviation of 1-2 µm often occurs. Thus, deterministic surface error corrections are required subsequently to reach the specifications, which employ cost-intensive and time-consuming sub-aperture finishing techniques.
In the case of small sized optics exhibiting diameters of less than 10 mm and freeform shape departures from best fit sphere of several tens of microns, grinding and polishing techniques come to geometrical limits due to the local slopes on the surface. Hence, small sized contact-less tools are preferred that are capable to meet the requirements on surface shape generation and smoothing. In the paper an alternative approach to freeform element fabrication is presented. The process chain comprises ultra-short pulse laser machining for shape generation by ablation on fused silica windows, followed by a plasma jet polishing step. This process chain is targeting the fabrication of individual small and medium-sized optics with diameters of 1 mm – 25 mm that exhibit steep surface slopes.

2 Experimental

The femtosecond laser system used for surface machining is a water-cooled CPA-2001 from CLARK-MXR. Pulses with duration of less than 150 fs at repetition rate of 1 kHz are emitted at a wavelength of 775 nm. For the experiments, the mean power has been adjusted by a polarization filter to approx. 50 mW. The laser system is equipped with fixed focussing optics. For substrate manipulation a 3-axis CNC motion system is employed. The plasma jet system utilized for polishing comprises a microwave driven plasma jet source fed by inert noble gases like argon or helium. Gas temperature in the plasma can reach up to 1500 °C, which is below the softening point of fused silica. The plasma jet is brought into contact to the surface and scans over the surface. The main advantage of plasma polishing technique over mechanical-abrasive polishing is that virtually no mechanical forces are applied to the surface [1]. Hence, the surface figure remains unchanged while roughness and sub-surface damage can be significantly reduced. All experiments have been conducted on optical grade fused silica windows. Surface form and roughness measurements have been performed by tactile profiler,

* Corresponding author: thomas.arnold@iom-leipzig.de

confocal microscopy, and white light interferometer microscopy (WLI), respectively.

3 Surface machining procedure

Laser machined areas on fused silica samples exhibiting constant depth have been plasma polished to evaluate the smoothing capability. Figure 1 shows cross section profiles after laser machining and after plasma polishing. Roughness values of 386 nm RMS before and 3.8 nm RMS after plasma polishing, respectively, have been measured indicating a significant smoothing effect. Remaining waviness in the spatial wavelength range of approx. 100 to 200 µm can be attributed in parts to laser power instabilities during the machining process. The WLI measurement shown in Figure 2 proves optical quality on the micro-roughness scale.

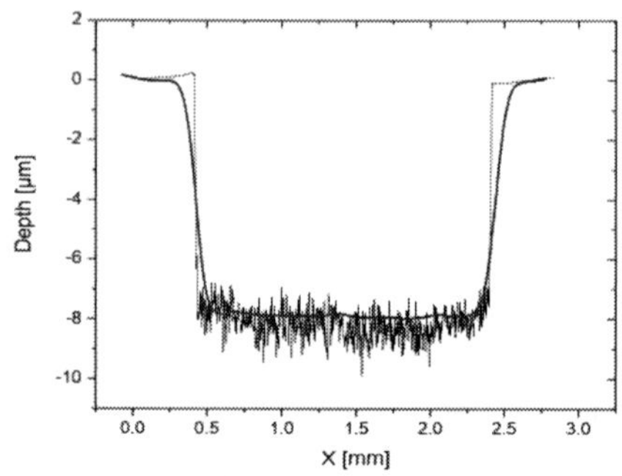

Fig. 1. Cross section profile of laser machined area before (black) and after plasma jet polishing (blue).

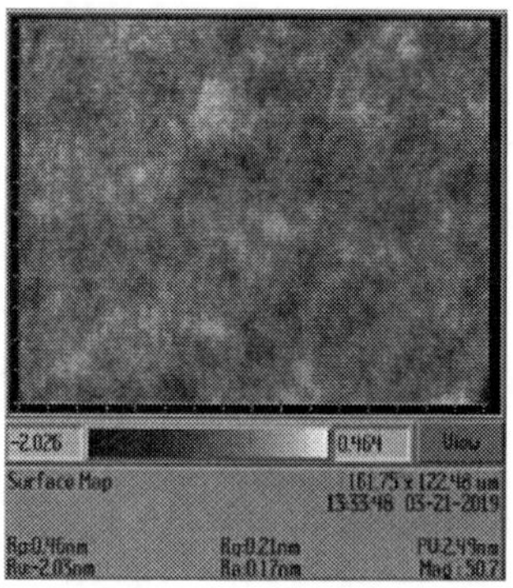

Fig. 2. Micro-roughness on plasma polished surface.

For the laser-based generation of a given surface shape, a dwell time controlled material removal is assumed. In order to determine laser machining performance with regard to ablation depth and material removal rate (MRR), a raster path motion scheme has been applied to generate area ablation of a constant depth D, where line feed was chosen as $dy = 10$ µm and scan speed was varied. As can be seen from Figure 3, the removal rates calculated

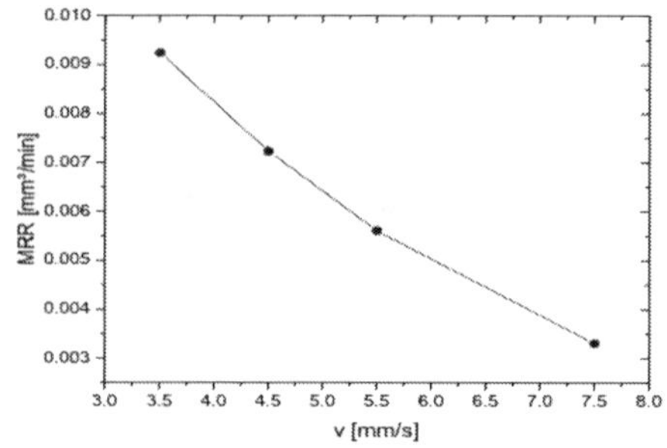

Fig. 3. Material removal rates determined area ablation depending on scan velocity.

by $MRR_{area} = v \cdot D \cdot dy$ show a clear dependence on scan speed. MRR data have been fitted to an analytical expression $f(v)$. By inversion of the resulting equation a function $v(D)$ was obtained that yields local speed depending on local depth distribution $D(x,y)$. This approach is a simplified variant of the deconvolution routines normally used in dwell time procedures, e.g. in plasma jet machining [2]. Unlike methods where extensive tool functions are used, which often have a Gaussian distribution, the tool function here is considered point-like. Figure 4 shows the targeted surface profile that was to be machined in x direction while the depth is constant in y direction. Additionally, the obtained profile after laser machining and after plasma polishing are depicted. Since the tactile profiler only measures the upper envelope of the rough surface, the profile seems to be a bit too flat. After plasma polishing the profile is close to the target shape. The results show that a combination of laser machining and plasma polishing may be beneficial for freeform generation, especially in the case where steep local slopes are required.

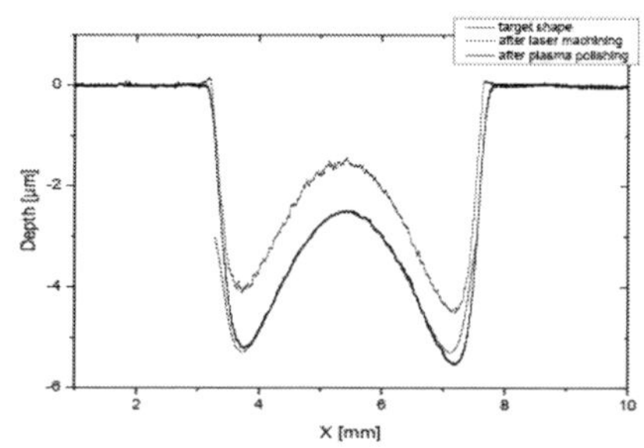

Fig. 4. Cross section profiles of targeted shape (red), after laser machining (black) and after plasma polishing (blue).

References

1. T. Arnold, G. Boehm, H. Paetzelt, J. Eur. Opt. Soc.-Rapid 11, 16002 (2016)
2. T. Arnold, G. Boehm, H. Paetzelt, Proc. of SPIE Vol. 9912, 99123N-1, 2016

Realization of depth reference samples with surfaces amplitudes between 0.1 nm and 5 nm

Annemarie Finzel, *Gregor* Dornberg, *Stephan* Görsch, *Martin* Mitzschke, *Jens* Bauer, and *Frank* Frost*

Leibniz Institute of Surface Engineering (IOM), D-04318 Leipzig, Germany

Abstract. A new approach for the realization of depth reference samples is presented. By a combination of photolithography, reactive ion beam etching, surface planarization with photoresists and a subsequent coating with non-transparent materials, defined sinusoidal surface profiles are generated which can be used as depth references for the comparison and calibration of different surface profile measurements. The smallest realized surface amplitudes are in the range of less than 0.1 nm.

1 Motivation

In the high-precision production of optical surfaces, roughness values in the range of 0.15 nm and below are now required. Such values are achieved by sophisticated polishing techniques, e. g. ion beam polishing. In addition, the fabrication of precision surface gratings with height modulation ranging from some nanometers down to sub-nanometer becomes more and more important, e. g. for diffraction gratings as used in vacuum–ultraviolet and soft x-ray spectral range.

In this context, the accurate and reliable measurement of corresponding surface profiles is of essential importance. Usually a high number of reference samples are commercially available which are used for the calibration/evaluation of surface profiling techniques as Scanning Force Microscopy (AFM) or White Light Interferometry (WLI). However, depth or roughness reference samples used for calibration of these methods have mostly relative high surface amplitude of some tens or some hundreds of nanometers. Ideally, the vertical dimensions of the features of the calibration samples should be in the same order of magnitude as the typically surface features to be measured.

In an earlier work [1] we could already show that reference samples with amplitudes down to 0.1 nm can be produced by the combination of lithography and reactive ion beam etching (RIBE). However, the process was relatively complex and time consuming, because a multi-layer system had to be patterned. Here a simpler process is proposed which allows to obtain sinusoidal surface profiles with low amplitude heights.

2 Experimental realization of depth reference samples

Figure 1 illustrates the process flow schematically. In the first part of the process, a lithographically defined resist mask is transferred by RIBE into a Si wafer, where the depth of the generated binary grating is adjusted by the etching time.

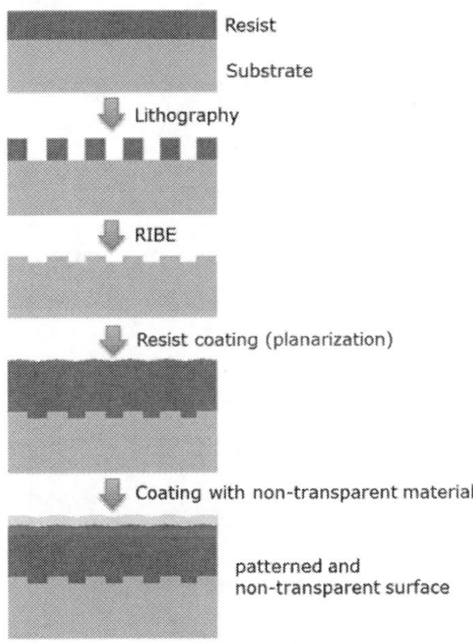

Fig. 1. Schematic process flow for the realization of the sinusoidal surface profiles.

In a second step a resist layer is applied with spin coating again which leads to a planarization of the surface and converts the previously binary profile into a sinusoidal profile. The surface planarization is caused by tension-driven flow of the viscous resist film [2]. The degree of planarization (reduction of surface amplitude) is determined both by the properties of the photoresist (viscosity, film thickness) and the properties of the initial Si surface profile (period, grating depth). In the final step, the wafer is coated with a thin layer (30 – 50 nm) of a non-

* Corresponding author: frank.frost@iom-leipzig.de

transparent material (Au, other metals, Si, ...). Here, magnetron sputtering was used which allows a conformal coating without changing the surface profile.

The Si wafers used had a diameter of 4 inches with regularly arranged fields of different periods and a lateral dimension of 15 mm x 15 mm being applied to the wafer. Other layouts are also possible.

3 Surface profile measurements using AFM and WLI

Figure 2 show AFM measurements of surface gratings with 4 µm period in (a) the Si surface after RIBE, (b) the resist coated planarized surface and (c) after final Au coating.

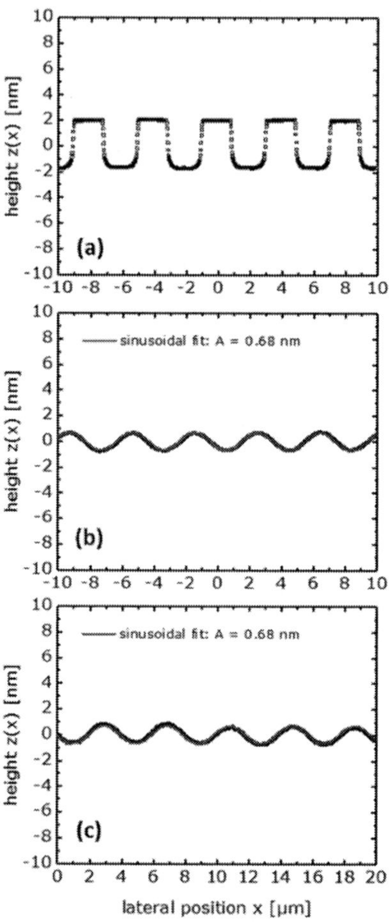

Fig. 2. AFM measurements of surface profiles with 4 µm period for (a) the initially binary grating in Si, (b) the sinusoidal profile after resist planarization and the final profile (c) after Au coating. The open symbols correspond to the measured AFM profile averaged over 256 scan lines, the red line is a sinusoidal fit to the measured profile.

The surface profile is not affected by the Au coating and the amplitude remains unchanged as indicated by the values of surface amplitudes (A = 0.68 nm) obtained from a sinusoidal fitting to the measured data.

Results for the WLI (50 x objective) measurement on a grating with 2 µm period are given in figure 3. The evaluation of the measurements results in a profile amplitude of (89.1 ± 4.4) pm. As described above, fields with a size of 15 mm x 15 mm were produced on the Si wafer. Measurements at different locations within a field show a typical standard deviation for the surface amplitudes of typically 5 pm, independent of the respective profile height. Nearly the same value was obtained from the corresponding AFM measurements.

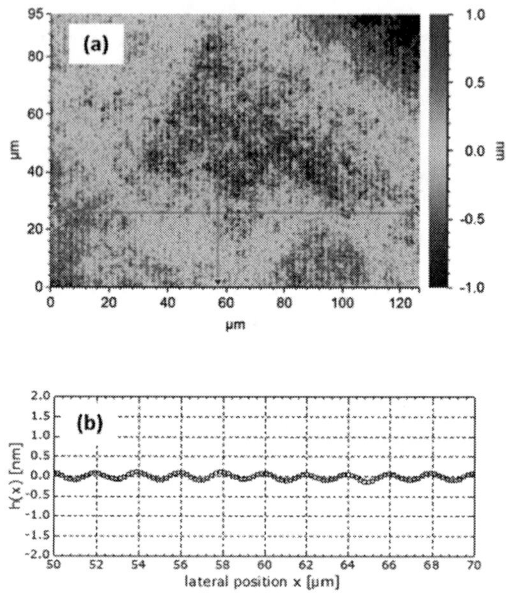

Fig. 3. WLI (50 x objective) measurements of Au coated surface profiles with 2 µm period: (a) measured surface topography, (b) averaged profile (open symbols), and sinusoidal fit to the measured profile (red line).

In conclusion a new approach for the realization of depth reference samples was presented. The adjustable profile amplitudes cover the range from approx. 5 nm to less than 0.1 nm. First evaluation measurements using AFM and WLI were presented.

The authors would like to thank Mrs. Katrin Ohndorf, Mrs. Petra Hertel and Mr. Toni Liebeskind for their technical support.

References

1. O. Faehnle, E. Langenbach, F. Zygalsky, F. Frost, R. Fechner, A. Schindler, M. Cumme, H. Biskup, C. Wünsche, R. Rascher, Proc. SPIE 9575, Optical Manufacturing and Testing XI, 957512 (2015)
2. Y. Li, H. Takino, F. Frost, Opt. Express. **25**, 7828 (2017)

https://doi.org/10.1051/epjconf/201921503005

Transient Dynamics Measurement of Light-material Interaction for Laser Optics

Zhichao Liu[1,*], *Feng* Geng[1], *Yaguo* Li[1], *Qinghua* Zhang[1], *Jian* Wang[1] and *Qiao* Xu[1]

[1]Fine Optics Engineering Research Center, Chengdu, 610041, China

Abstract. Laser damage performance is an important parameter to evaluate the quality of high power laser optical components. Understanding the physical mechanism of the interaction between laser and optical material is of great significance for improving the laser damage resistance of optical elements. For nanosecond pulsed laser, the time duration of laser-induced damage is ultra-short, and the measurement of related physical quantities is very difficult. Therefore, the Time Resolved Pump and Probe (TRPP) technique can effectively solve this problem. TRPP shadow imaging can obtain the transient dynamic information in range of picosecond to microsecond, including the plasma formation in the early stage, the shock wave propagation, the crack expanding, material ejection and other phenomena. Moreover, the ultra-fast physical quantities can be accurately measured by employing two detective beams with perpendicular polarization directions, for example S and P polarization light. All those information provides effective experimental data to reveal the physical mechanism of damage process.

1 Introduction

The interaction mechanism between high-power laser and optical material in nanosecond region is a quite important issue for various researches, such as laser assisted coating, laser ablation and laser induced damage. For general broadband transparent dielectric materials [1], the laser energy density required for inducing ionization breakdown usually exceeds to 10^{11} W/cm^2. The temperature of the irradiated area increases to 1 eV~10 eV and the local pressure exceeds to 10 GPa~100 GPa [2] in an ultra-fast time. A variety of physical phenomena will be involved in this process, such as the shockwave propagation, crack expanding, stress conduction and light absorption evolution. They are important to reveal the interaction mechanism of laser and material in the early time. Characterization of the physical parameters in this region will be useful to build theoretical model and understand the mechanism of light-matter interaction.

The Time Resolved Pump and Probe (TRPP) technique can effectively solve this problem. TRPP shadow imaging can obtain the transient dynamic information in range from ps to ms. Moreover, the ultra-fast physical quantities can be accurately measured by employing two detective beams with perpendicular polarization directions. In this paper, we introduce the TRPP device and its freatures. The plasma formation in the early stage, shock wave propagation, crack expanding and material ejection are investigated as the typical application of TRPP shadow imging technique in transient measurement field.

2 Principle and device

The principle of TRPP technology is illuminated in Fig. 1. It contains pump laser, probe laser, digital delay device, optical delay line and shadow imaging system. The pump laser provides a nanosecond pulse to irradiate the sample surface and creates damage region. In the meanwhile, a picosecond pulse functioning as the probe laser goes through this region to take the transient image, grabbing a time-slice. Finally, the transient image is recorded and extracted by the shadow imaging system. The different time-slices of the whole damage event can be grabbed by shifting the delay between the pump and the probe pulse. In order to measure the transient parameter quantitatively, we need to grab two slices for one damage event. To achieve this purpose, the probe laser is split into two beams with orthogonal polarization directions, S light and P light, by use of a polarized beam splitter (see PBS in Fig. 1). The relative delay between them is controlled by an optical delay line, as denoted by the dotted line frame in Fig. 1. After that, the two beams are combined coaxially by another PBS and go through the damage region together. Passing the damage region, the beams are split to S and P lights again by the third PBS. Two microscopes are deployed to grab the transient images of S and P lights as two time-slices.

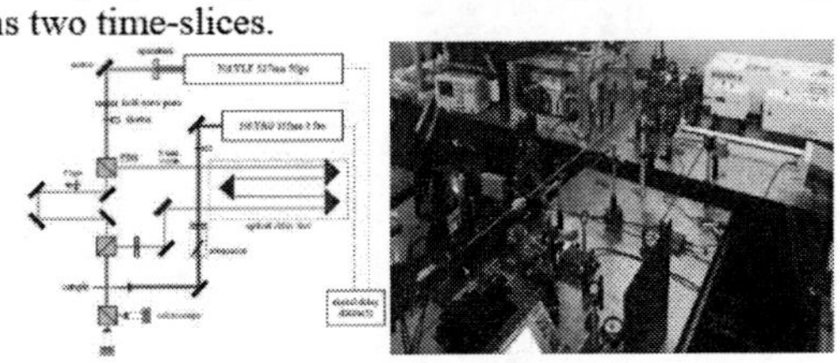

Fig.1 Schematic diagram of the TRPP setup.

[*] Corresponding author: zcliu44@163.com

 https://doi.org/10.1051/epjconf/201921503005

3 Results and discuss

3.1 Plasmas formation

The avalanche ionization region continues expanding to explosion and then induces the shockwave. The transient images of this early stage were caught by the TRPP method shown in Fig. 2. The shadow cluster is the ionization phenomenon, for example, pointed by the white arrow in Fig. 2(b). Once the probe light is absorbed by plasma, the dark image will appear. Therefore, by measuring the onset time, position and size of shadow cluster, we can obtain the evolution of plasmas region.

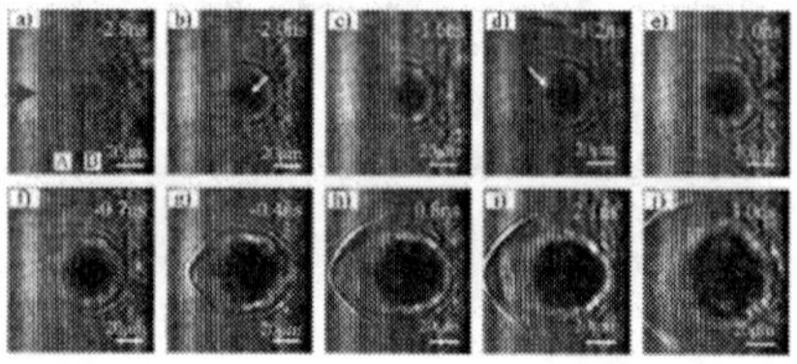

Fig. 2. The propagation of plasma and shockwave in pump laser pulse duration from -2.8 ns to 3.0 ns. Red dotted line means the interface of air side (signed as A) and bulk side (signed as B). Pump laser comes along with red arrow.

3.2 Shockwave propagation

The Fig. 2 shows the image series of shockwave propagation in the air side under nanosecond laser pumping. Each sub-image is caught with a certain delay. The pump laser comes along with the red arrow. By using P/S light method, the velocity of shock wave can be achieved as shown in Fig. 3. Its decay rule obeys the Taylor-Sedov's point explosion model [3].

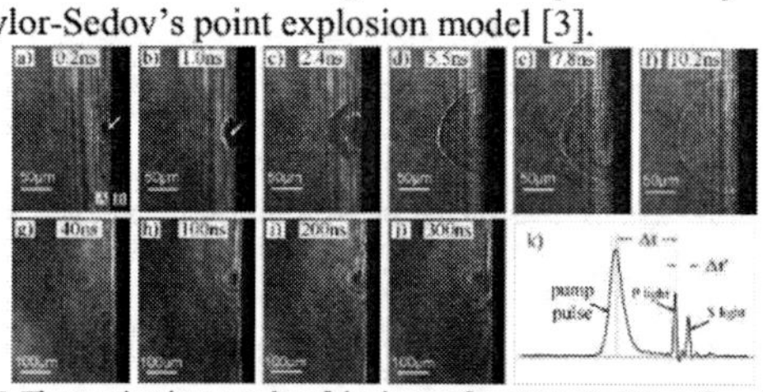

Fig. 3. The transient image series of shockwave from 0.2 ns to 300 ns (from (a) to (j)) and the schematic diagram of S-P light method (k). Pump laser comes along with red arrow and the white arrow means the plasma region.

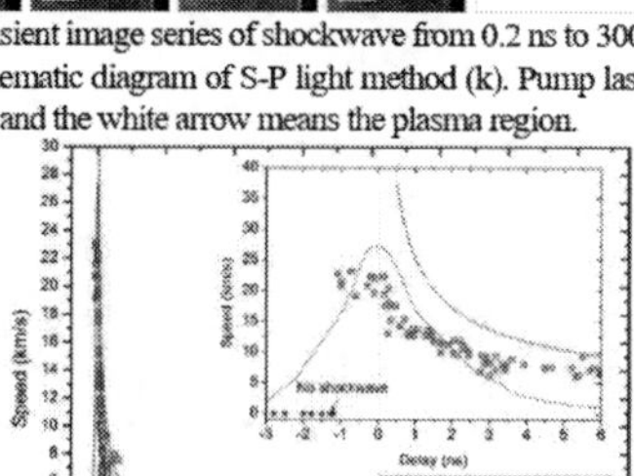

Fig. 4. The relationship between shockwave speed and delay from -1.2ns to 300ns under 0.77mJ pump energy. The blue dotted line is the fitting curve of shock wave speed along with delay. And the grey curve is the contour of pump laser duration.

3.3 Crack expanding

Use Vickers indenter to suppress artificial cracks on the surface of KDP crystal. The dynamic evolution of indentation pits and cracks towards damage under laser irradiation was observed, see Fig. 5. The upper images were taken under bright field and lower ones were under dark field. The results indicates a clear correlation between the position of plasma absorption cluster and the crater crack.

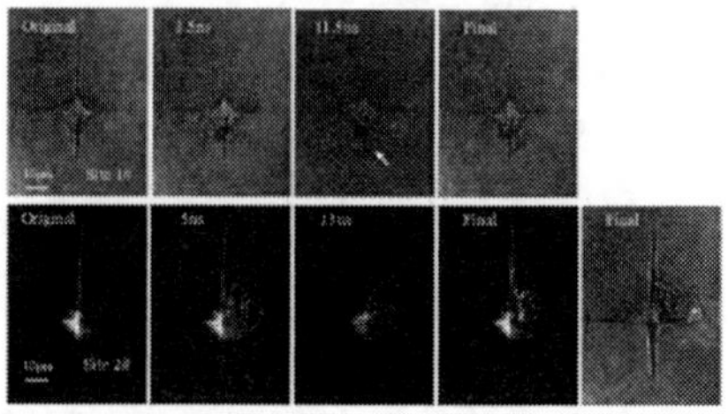

Fig. 5. Dynamics of the cracks expanding under laser irradiation.

3.4 Material ejection

By using TRPP technology, the material ejection process can be observed, see Fig. 6. Some important physical quantities such as ejection angle, velocity and onset time can be obtained, which can provide experimental support for the establishing the mechanical fracture model.

-The ejection angle of the rear surface is larger than that of the front surface.

-Two stages: For delay less than ~1μs, melted and spherical particles. For delay more than 1μs, irregular fragments with larger size and lower speed.

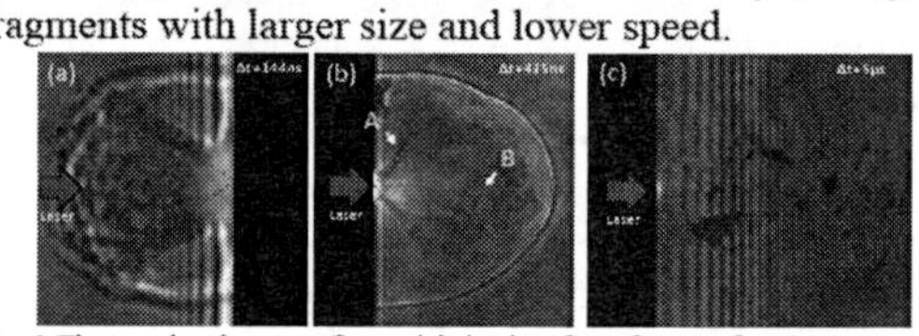

Fig. 6. The transient images of material ejection, from front surface (a), rear surface (b) and the debris ejection for longer delay.

4 Conclusion

In summary, the TRPP technique discrribed in this paper can be used to reveal the transient dynamics process of light-matter, espectically for high power laser-induced damge. Some most important physical parameters, such as plasmas size, shockwave speed and material ejection angle, could be measured quantitatively. Thanks to the TRPP technique, we can get the rule of ultra-fast process and to understand the mechanism of laser-induced damage.

References

1. B. C. Stuart, M. D. Feit, S. Herman, A. M. Rubenchik, B. W. Shore, and M. D. Perry, Phys. Rev. B **53**, 1749 (1996)
2. Carr, C., Radousky, H., Rubenchik, A., Feit, M., and Demos, S., Phys. Rev. Lett. **92**, 087401 (2004)
3. Geoffrey Taylor, Proc. Royal Sco. London, Ser. **A201** (1950)

EPJ Web of Conferences **215**, 04001 (2019) https://doi.org/10.1051/epjconf/201921504001
EOS Optical Technologies

Economical rapid-prototyping of aspherical lenses

Angelina Müller[1], *Matthias C.* Wapler[1], and *Ulrike* Wallrabe[1,*]

[1]University of Freiburg, IMTEK - Department of Microsystems Engineering, Laboratory for Microactuators, Freiburg, Germany

Abstract. We present a rapid-prototyping process to fabricate aspherical lens arrays based on surface deformation due to thermal expansion of PDMS. Using laser-structuring and molding in combination with an FEM-based shape optimization, we were able to design, fabricate and characterize different micro-lens arrays. This fabrication process can be used for almost any kind of arbitrary lens shape, which allows for a large design freedom for micro lenses.

1 Introduction

Freeform micro lenses and micro lens arrays have many applications, e.g. in imaging systems, 3D displays, optical sensors or light sources [1]. Cost, design freedom and quality are the essential key figures. Some manufacturing methods have the disadvantage of being very expensive and time-consuming, e.g., precision machining technologies, two-photon polymerization techniques or grayscale lithography. Other processes including inkjet printing or reflow processes have limitations in the lens geometry. Our rapid-prototyping process, however, can produce nearly any lens shape, allowing a high degree of freedom in the design of micro-optical lenses. In particular, it is cost-effective also for small quantities.

2 Design and fabrication

Our fabrication method is based on the surface deformation due to thermal expansion of a soft polymer, polydimethylsiloxane (PDMS), in a cavity, first described by Brunne et al. [2]. The fabrication consist of two molding steps as illustrated in Fig 1. First, we laser-structure the initial cavity into a thick silicon substrate. Then, we fill liquid PDMS (Sylgard 182 from DowSil) into the cavity, cover it with a silane-coated glass slide and cure it in a mechanical press at a pressure of approx. 10 MPa and a temperature of 120 °C. Before curing, the liquid PDMS expands during heating, such that excess material escapes from the cavity via spacers (see image of step (2)). After approx. 40 min the PDMS is polymerized, we remove the slide and the material shrinks during cooling. This results in a small but smooth surface deformation, which is generally dependent on the depth and shape of the initial cavity, the polymerization shrinkage, the coefficient of thermal expansion and the Poisson's ratio. The deformed surface acts as a mold in the second molding step where we fill the mold with a UV-curing adhesive, cover it with a glass substrate and cured it with UV exposure. We were able to determine the shape of the initial cavity for almost any lens shape using an FEM shape optimization, which needs the desired PDMS curing temperature and material

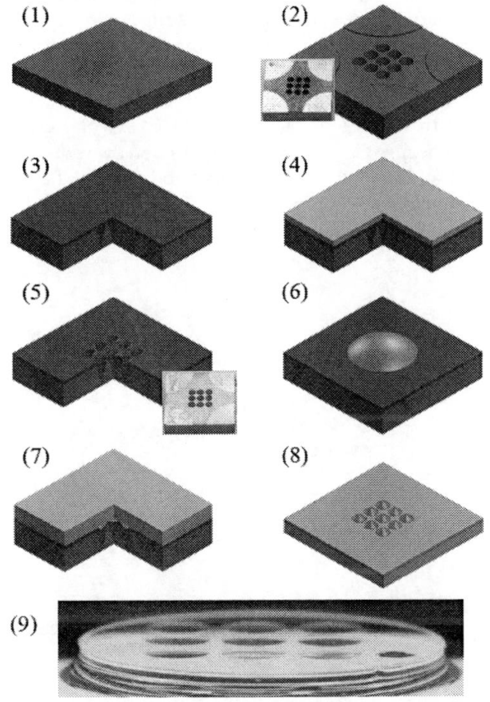

Fig. 1. Fabrication process of the transmissive aspherical optics. A funnel-shaped cavity is laser-structured into a thick silicon substrate (1 and 2). The uncured liquid PDMS is filled into the cavity (3), sealed with a glass slide (4) and heated under pressure. The still liquid polymer expands due to the elevated temperature and is pressed out of the mold. After polymerization, the device is cooled down, and the cured polymer shrinks. Small surface deformations form the lens molds (5). A UV-curable adhesive is dispensed on top of the lens molds (6), and a glass slide is pressed on (7). The device is then cured with UV-exposure, and the glass slide with the cured UV-adhesive is removed (8).

* Corresponding author: wallrabe@imtek.uni-freiburg.de

parameters. For this purpose, we have recently developed a new, accurate and quick method to determine the Poisson's ratio and the coefficient of thermal expansion of PDMS [3].

3 Characterization and results

We demonstrate our process with a 3 x 3 microlens (axicon) array with an element diameter of 1 mm and a conical lens shape with 174° apex angle.

For the characterization of our lens array as shown in Fig. 2, we measured the surface quality of each lens with a high-precision optical profilometer. We determined the deviation of the lens shape from the intended shape (simulated target function f_{target}) for three different positions (Fig. 3), where we can observe smaller deviations. Lens P5 deviates strongly from the intended lens shape in the center, which suggests some particle, air bubble or void in the PDMS during the fabrication process. Furthermore, we determined the surface error of the three lens positions as a deviation from rotational symmetry and found a surface error of 113 ± 13 nm (Tab. 1). In addition to the surface measurement, we verified the optical performance by illuminating the axicons with a collimated laser beam of λ = 473 nm and measured the resulting beam profile. In Fig. 4, we see a very good quality of the Bessel beam along with little distortion as expected from [4] with a reconstructed wavefront error of around $\lambda/6$.

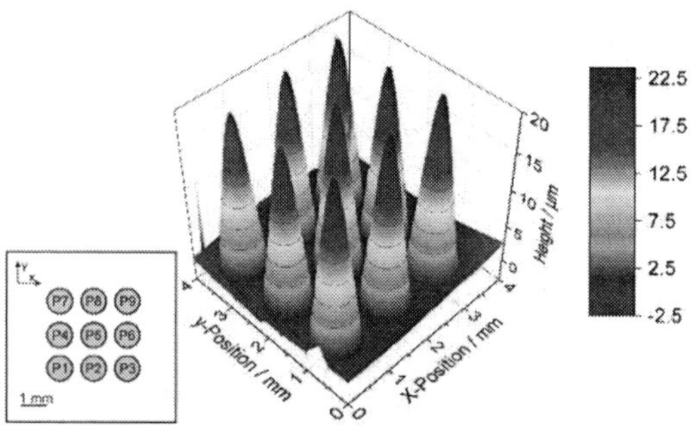

Fig. 2. Measured three-dimensional surface data of a fabricated 174° axicon array with indications of the lens positions.

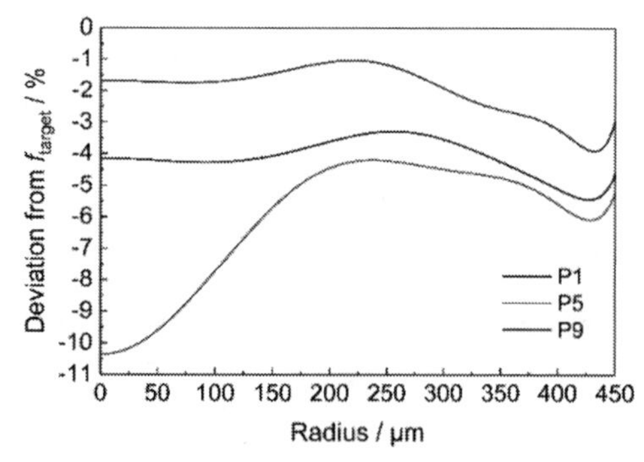

Fig. 3. Deviation of the measured surface data from the simulated target function for three different measured positions.

4 Conclusions

We demonstrate a new rapid-prototyping process, which can produce lenses of almost arbitrary shapes. Combing precise FEM simulations with laser-structuring using only a marking laser and a final molding process, we were able to fabricate a 174° axicon array with a low mean rms surface deviation of 113 ± 13 nm. This value is comparable with other state-of-the-art technologies for the production of microlenses [5]. In addition, we were able to generate a Bessel beam profile, which showed only small wavefront errors in the range of $\lambda/6$, indicating high optical quality and the required high rotational symmetry. Our new rapid prototyping process therefore permits the optically precise production of strongly aspherical optics in the mm range and is also array-compatible.

Tab 1. Measured rms surface deviation at three different lens positions.

Lens position	Surface error
P1	99 nm
P5	118 nm
P9	123 nm
Mean value	113 nm
Standard deviation	±13 nm

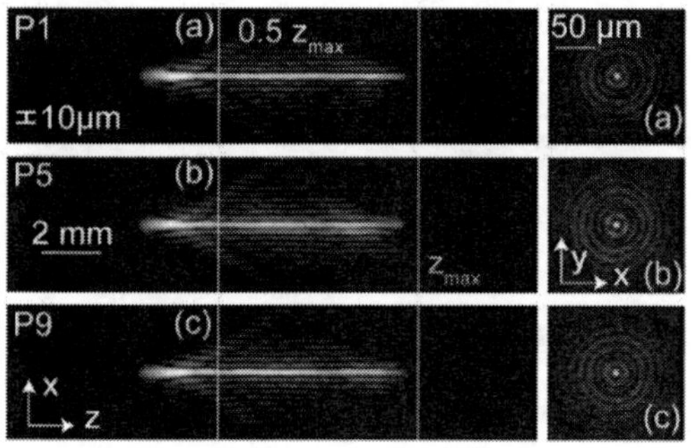

Fig. 4. Measured intensity profile of the Bessel beam (logarithmically scaled) for three different positions (left) with cross-sections at $z_{max}/2$ (right).

This work was supported by the BrainLinks-BrainTools Cluster of Excellence funded by the German Research Foundation (DFG, grant no. EXC 1086).

References

1. Y. Hu, Y. Xiong, X. Chen, H. Bai, Y. Tian, G. Liu, App. Opt., **57**, 15, pp. 4277-4282 (2018)
2. J. Brunne, U. Wallrabe, Opt. Letters, **38**, 11 pp. 1939-1941 (2013)
3. A. Müller, M.C. Wapler, U. Wallrabe, Soft Matter, **15**, 4 pp. 779-784 (2019)
4. A. Müller, M.C. Wapler, U.T. Schwarz, M. Reisacher, K. Holc, O. Ambacher, U. Wallrabe, Opt. Express, **24**,15 pp. 17433-17452 (2016)
5. H. Ottevaere et al., J. Opt. A, **8**, 7 pp. 407-429 (2006)

Reactive ion beam etching – based finishing of optical aluminium surfaces

Melanie Ulitschka[1,*], *Jens* Bauer[1], *Frank* Frost[1], and *Thomas* Arnold[1,2]

[1]Leibniz Institute of Surface Engineering (IOM), D-04318 Leipzig, Germany
[2]Technische Universität Dresden, Institute of Manufacturing Science and Engineering, D-01069 Dresden, Germany

Abstract. Reactively driven low-energy ion beam machining is a widely used finishing technique for optical aluminium surfaces. Direct RIBE machining with oxygen or nitrogen process gas permits figure error correction of diversely shaped aluminium optics while preserving the initial roughness up to 1 µm etching depth. Those optical surfaces are appropriate for applications in the infrared spectral range. For increasing demands on optical surfaces for short - wavelength applications in the visible and ultraviolet spectral range ion beam polishing techniques with the aid of a sacrificial layer are a promising process route. Turning marks situated in the high-to –mid spatial frequency range are remarkably reduced while preserving the initial microroughness resulting in an enhancement of the full range roughness.

1 Introduction

For increasing demands on high-performance optical systems, scattering effects resulting from surface errors are necessary to reduce.

Technical aluminium alloy materials as Al6061 or Al905 are common mirror construction materials. For fabrication of optical aluminium surfaces conforming to the requirements in the infrared (IR) spectral range, usually single-point diamond turning (SPDT) is used [1, 2]. Periodic groove pattern microstructures (turning marks) are left on the surface having similar effects of a diffraction grating (Fig. 1).

However, the demands on optical surface quality for applications in the short-wavelength visible (VIS) and ultraviolet (UV) spectral range increase immensely. In particular, improved surface qualities with roughness values of 1 nm root mean square (rms) down to 0.3 nm are required [3].

The surface errors to be corrected are specified by the spatial frequency range they emerge [4, 5]. Depending on the lateral dimension, the surface errors are divided into low-spatial frequency (LSF), mid-spatial frequency (MSF) and high-spatial frequency (HSF) errors.

Deviations of the surface form or shape (i.e. surface figure errors) are LSF errors. On the contrary, HSF errors are microroughness or surface finish errors causing diffusive shine. MSF errors are situated in between. Turning marks are typical feature types situated in the mid - to - high spatial frequency range of waviness and roughness (Fig. 1).

The surface errors to be corrected are on the nanometer and sub-nanometer scale. Thus ion beam technologies driven by reactive process control are a promising process route to improve the optical performance since these slow surface machining techniques offer a high degree of precision.

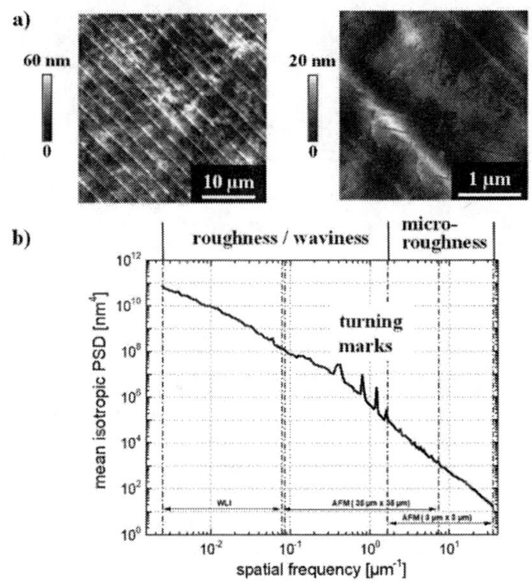

Fig. 1. Optical aluminium surface of Al905 after SPDT measured with a) Atomic force microscopy (AFM) with scan sizes of 35 µm x 35 µm (top) and 3 µm x 3 µm (bottom). b) Calculated power spectral density (PSD) function based on the AFM measurements depicted in a) and WLI measurements (1230 µm x 925 µm).

* Corresponding author: melanie.ulitschka@iom-leipzig.de

2 Deterministic ion beam machining

Ion beam etching technologies driven by reactive process control (N_2, O_2) allows figure error correction of optical aluminium surfaces up to 1 µm machining depth while preserving the initial surface roughness (Fig. 2b-d).

The interaction of a low- energy ion beam with the aluminium surface suggest the formation of a quasi-stationary surface passivation layer (Fig. 2a). The erosion process is based on pure physical sputtering. Hence, the etch rate is strongly dependent on the ion energy and the ion incidence angle [6, 7].

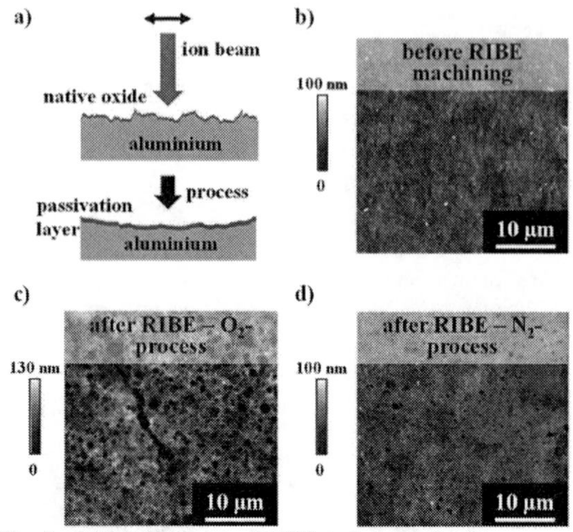

Fig. 2. a) Model scheme of RIBE techniques for figure error correction via RIBE machining with oxygen or nitrogen process gas. AFM topography images (35 µm x 35 µm) of Al905 samples b) before RIBE machining and after ion beam machining with c) oxygen and d) nitrogen process gas.

3 Ion beam planarization

For ion beam planarization technique, a photoresist layer is deposited by conventional photolithography exhibiting an ultra-smooth surface. The technological thick photoresist layer fully embeds the surface roughness features of the initial aluminium surface. Removing the resist by reactive ion beam processing, the ultra-smooth surface topography of the resist is transferred into the aluminium substrate (Fig. 3a).

The selectivity, defined as the ratio of the etching rates of aluminium substrate and photoresist layer, significantly affects the planarization process. For optimized surface transfer a selectivity of 1:1 needs to be reached which can be achieved by accurately adjustment of the ion incidence angle and process gas composition.

Photoresist thermal pre-treatment was optimized to allow roughness preservation during RIBE machining and steady-state material removal rate.

During ion beam planarization the turning marks present on the initial aluminium surface are remarkably reduced in height while preserving the initial microroughness. Consequently, the full range roughness is strongly reduced (Fig. 3b-d).

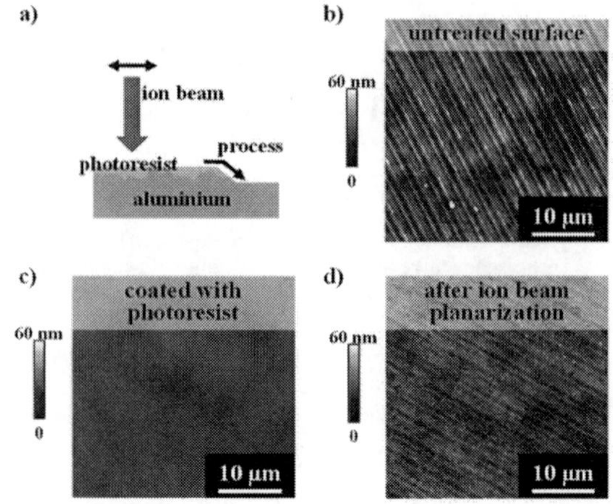

Fig. 3. a) Model scheme of RIBE techniques for planarization with the aid of a photoresist layer. AFM topography images (35 µm x 35 µm) of b) the untreated Al905 surface, b) coated with a photoresist layer and c) after ion beam planarization.

The authors are very grateful to Mr. T. Liebeskind for sample preparation and chemical sample treatments. Financial support by the German Federal Ministry of Education and Research (BMBF) within the framework of the InnoProfile Transfer initiative 03IPT706X 'Ultra-precision manufacturing using atomic particle beams' is gratefully acknowledged.

References

1. R. Steinkopf, A. Gebhardt, S. Scheiding, M. Rohde, O. Stenzel, S. Gliech, V. Giggel, H. Löscher, G. Ullrich, P. Rucks, A. Duparre, S. Risse, R. Eberhardt, A. Tünnermann, *Proc. SPIE* **7102**, *Optical Fabrication, Testing, and Metrology III*, 71020C (2008)
2. G. P. H. Gubbels, B. W. H. Venrooy, R. Henselmans, *Proc. SPIE* **7426**, *Optical Manufacturing and Testing VIII*, 742607 (2009)
3. M. Weiser, Nucl. Instruments Methods Phys. Res. Sect. B Beam Interact. With Mater. Atoms, **267**, 1390-1393 (2009)
4. J. E. Harvey, A. Kotha, *Proc. SPIE - The International Society for Optical Engineering* **2576**, 155-174 (1995)
5. J. M. Bennett, Meas. Sci. Technol. **3**, 1119-1127 (1992)
6. J. Bauer, M. Ulitschka, F. Frost, T. Arnold, *EOS Opt. Technol. / 5th Conf. Manuf. Opt. Syst.*, (European Optical Society, Munich / Germany 2017)
7. J. Bauer, F. Frost, T. Arnold, J. Phys. D. Appl. Phys. **50**, 85101 (2017)

EPJ Web of Conferences **215**, 04003 (2019) https://doi.org/10.1051/epjconf/201921504003
EOS Optical Technologies

Splicing fluoride glass and silica optical fibers

Solenn Cozic[1,*], *Simon* Boivinet[2], *Christophe* Pierre[2], *Johan* Boulet[2], *Samuel* Poulain[1] and *Marcel* Poulain[1,3]

[1]Le Verre Fluoré, Rue Gabriel Voisin, Campus Kerlann 35170 Bruz, France
[2]Alphanov, Rue François Mitterrand, 33400 Talence, France
[3]Institut Sciences Chimiques de Rennes, Campus Beaulieu, 35042 Rennes Cedex, France

Abstract. Splicing fluoride glass fibers and silica fibers is a critical point for manufacturing all fibered laser modules. As these materials are extremely different, various problems must be considered: thermal, expansion, mechanical, chemical. Reliability and power handling make priority concerns. We report splices made on a 200/220 multimode silica fiber and a double clad 15/250/290 ZBLAN fiber. Splices are proof tested at 300 g tensile strength. No damage is observed after thermal cycling from -30 °C to 85 °C, at 40 % RH during 24 hours. Typical optical splice loss is about 0.2 dB. They withstand 220 W input power at 976 nm without any damage and drastic temperature increasing.

1 Introduction

Fluoride glass fibers are used for the fabrication of various fiber lasers operating in the mid infrared spectrum (> 2 μm) [1-5] and for supercontinuum generation from 1 to 5 μm [6,7]. The industrial manufacturing of such laser sources requires splicing silica fibers and fluorides fibers to handle the injection of the pump laser. Thermal splicing has been successfully implemented by several research teams [1, 2, 5]. Reproducibility and reliability of such splices have not been investigated.

This operation raises a number of technical problems in relation to the physical and chemical characteristics of the two materials. A major difference concerns the thermal expansion coefficients: 5.5 10^{-7} K^{-1} for SiO_2 and 170 10^{-7} K^{-1} for ZBLAN. As a consequence, large mechanical stress will occur at cooling and for any temperature change. Another challenge arises from the steep viscosity change with temperature for fluoride glasses: glass may flow, which leads to the geometric deformation of the fiber. It must be emphasized that the characteristic temperatures of both glasses are extremely different: softening temperature of ZBLAN is below 300 °C while it is above 1000 °C for vitreous silica. At this temperature fluoride glass decomposes and releases gaseous species. A last problem is the control of the chemical processes at the surface of fluoride fibers. Depending on temperature, silica reacts with fluoride glass melts to produce gaseous silicon tetrafluoride SiF_4 according to the chemical equation:

$$SiO_2 + 4\,F^- \;=>\; SiF_4 + 2\,O^{=}$$

Thus temperature appears as a critical parameter. Below glass transition temperature Tg, fiber is in the solid state. Beyond Tg, glass enters the liquid state. It is a necessary condition to create the chemical bonding required for a strong splice. Then the temperature range for which splicing is possible may be narrow.

Insofar as developments focus on high power fiber lasers, one must ensure that pump lasers do not induce damage in fiber core and cladding. Previous experiments have already demonstrated that pump power in excess of 200 W could be injected in the first cladding of a double clad ZBLAN fiber [3]. Therefore, development of splices handling such high average power is required.

Here, we report the new developments of splices between fluoride glass fibers and silica fibers exhibiting good mechanical and optical performances.

2 Experimental

2.1 Fibers

The fibers used in this work are: 1) a 200/220 0.22 NA silica fiber from Thorlabs and 2) a 14/250/290 double clad ZBLAN fiber with a second cladding made of low refractive index polymer.

2.2 Equipment and processing

The experiments were conducted at ALPHANOV laboratories, using a water cooled, 976 nm laser diode

* Corresponding author: solenn.cozic@leverrefluore.eu

 https://doi.org/10.1051/epjconf/201921504003

from Dilas, pigtailed to a 200 μm silica fiber with 0.22 NA, an injection bench with two aspheric lenses, a Pl160 thermal camera to monitor the splice temperature, and Ophir air cooled powermeter handling 1.1 kW. After stripping the polymeric coating, fibers were prepared using a LDC401-A Vytran cleaving machine and a GPX3400 Vytran splicing machine. The two fibers set in the splicer and heated to reach the softening temperature of the fluoride glass. Then they were put into contact and the pressure was released while they were slowly cooled. In a last step, the splice was coated with a low refractive index, UV curable resin. 5 splices were made in this way.

3 Results

The first measurements have shown that splices could withstand a 450 g tensile strength. Then a proof test of 300 g during 30 s was implemented for the splices used for this study. A splice loss of 0.2 dB has been measured at 1550 nm. The picture of a typical splice is given in figure 1. The small difference in fiber diameter may be seen and also the overlap of the fluoride fiber over silica.

Thermal cycling at 40 % RH was implemented for 24 hours, between -30 °C and 85 °C. It does not alter the quality of the splice

Coupling of CW 976 nm laser diodes in the first fluoride cladding was made at increasing power. No damage occurred on the splice, even at a input power higher than 200 W. The splice temperature has been monitored while the injected average power is increased. At 220 W input power, the splice temperature is only increased to 40 °C without any cooling packaging: indeed the splices were only laid on the table. This feature confirms that the splice loss is very low. Note that we injected only 220 W onto the silica fiber due to no-optimal coupling, leading to a very high temperature increase at the entrance of the silica fiber. This issue could be solved by the splice between the diode output fiber and silica fiber.

Figure 1. Splice of SiO_2 and ZBLAN fibers.

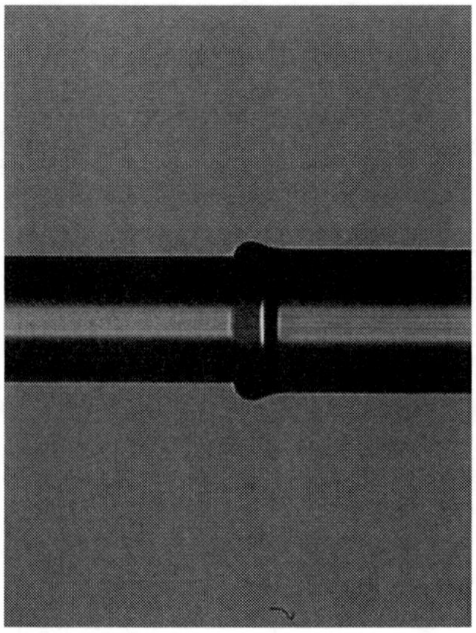

4 Conclusion

Silica fibers and fluoride glass fibers were successfully spliced and characterized. These splices withstand significant tensile strength (> 300 g), large temperature variations (-30 °C to 85 °C) during 24 hours, and high power laser of 220 Watts CW without excessive temperature increase.

These features make possible to manufacture compact, all fibered and reliable high power fiber lasers in the mid-IR.

References

1. H Okamoto, K. Kasuga, and Y. Kubota, Opt. Lett. 36, 1470 (2011)
2. V. Fortin, M. Bernier, R. Vallée, Opt. Lett, **40**, 2882 (2015)
3. Y. O. Aydin, V. Fortin, R. Vallée, M. Bernier, Opt. Lett., **43**, 4542 (2018).
4. L. Yang, B. Zhang, D. Jin, T. Wu, X. He, Y. Zhao, J. Hou, Opt. Lett., **43**, 5206 (2018).
5. V. Fortin, F. Jobin, M. Larose, M. Bernier, and R. Vallée, Opt. Lett., **44**, 491(2019).
6. J. C. Gauthier, V. Fortin, J. Y. Carrée, S. Poulain, M. Poulain, R. Vallée, M. Bernier, Opt. Lett., **41**, 1756 (2016)
7. L. Yang, B. Zhang, D. Jin, T. Wu, X. He, Y. Zhao, and J. Hou, Opt.Lett **43**, 5206, (2018)

Testing the transmitted wavefront of large aperture long-focal-length lens using a multizone computer-generated hologram

Jian-Peng Cui*, *Ning* Zhang, *Jie* Liu, *Di-Long* Wu, *Hua* Xu, *Ding-yao* Yan and *Ping* Ma

Chengdu Fine Optical Engineering Research Center, 610041 Chengdu, China

Abstract. A method for testing the transmitted wavefront of large aperture long-focal-length lens with a multizone computer-generated hologram (CGH) is proposed. The multizone CGH has 5 zones: one main zone for the null testing of long-focal-length lens and four auxiliary zones for the pre-alignment of measured lens. Both 1st order wavefront and 0th order wavefront of CGH are measured, and 0th order wavefront is used to calibrate the substrate error. To verify this test approach, a 450mm×450mm multizone CGH is designed and fabricated for testing the spatial filter lens. Experiments and error analysis are carried out. The results show that the desired precision can be reached with use of CGH.

1 Introduction

Large aperture long-focal-length lens has been widely used as the spatial filter lens in high power laser system, such as National Ignition Facility in Unite States, Laser Mégajoule in France, Gekko XII in Japan. Transmitted wavefront is an important specification for long-focal-length lens and will affect the filtering effect and beam quality directly. One method for testing the transmitted wavefront involves using a large convex mirror [1-2]. This method has a simple test configuration and short optical path. However, measurements of large convex mirrors are notoriously difficult because they require flat interferometer and auxiliary optics that are larger than the surfaces being tested, which results in a high cost.

Current developments in diffractive optics and lithography technology make the use of large aperture computer-generated hologram (CGH) an attractive alternative [3-4]. Therefore, in this paper, we propose to measure the transmitted wavefront of long-focal-length lens by using a reflective multizone CGH. Compared with the convex mirror, CGH has a large advantage in cost, because it is cheaper to manufacture a diffractive element than convex mirrors of equivalent size. Moreover, the design of CGH is very flexible, one multizone CGH can provide the wavefront correction for null testing and pre-alignment of the lens under test simultaneously, which makes the alignment of optical elements easier.

2 Measurement principle

Figure 1 shows the CGH test configuration for long-focal-length lens. A large aperture Fizeau interferometer and a reflective multizone CGH is used to measure the transmitted wavefront. The multizone CGH consists of one main zone and four auxiliary zones. The main zone is located at the center of the substrate. In this zone, testing CGH with circular lines is etched to emulate, in 1st order (or higher order), the reflective properties of the convex mirror. Auxiliary zones are four square zones and located around the main zone. In auxiliary zones, four beam projection CGHs are etched for the prealignment of the measured lens by projecting four marks around the edges of the measured lens.

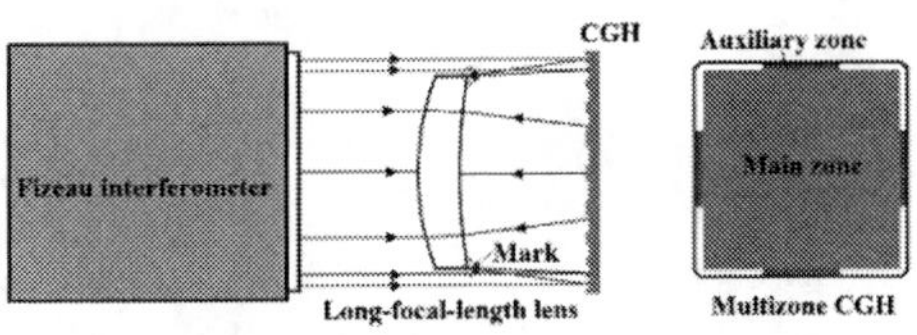

Fig. 1. Schematic drawing of the CGH test configuration for long-focal-length lens.

As we know, CGH substrate figure error is usually the primary error source in the CGH measurement. To calibrate the CGH substrate figure error, two measurements are made: 0th order measurement and 1st order measurement. Through measuring the 0th order diffraction wavefront of the CGH, we can obtain the substrate figure error and back it out from 1st order one.

3 Experiments and results

3.1. 0th and 1st order measurements

To verify the feasibility of this test approach, we designed a multizone CGH for testing the spatial filter lens. This spatial filter lens has a size of 430mm×430mm and thickness of 46.5mm. The effective focal length f is 31250mm (@λ=1053nm) and the vertex radii of curvature are R1=9377.868mm (CX) and R2=28133.886mm (CC), respectively. In 410mm×410mm clear aperture, the requirement of transmitted wavefront is PV<λ/3

* Corresponding author: 547150445@qq.com

(@λ=632.8nm). The multizone CGH is fabricated on a 450mm×450mm×70mm fused silica substrate and the substrate is specified to have a figure of PV~1/10λ. The testing CGH is 430mm×430mm and the minimum line spacing is 32.1μm. In our experiment setup, the large aperture interferometer is a commercialized Fizeau interferometer INF600-LP produced by Tyggo, which has a test aperture of 610mm, as shown in Fig. 2.

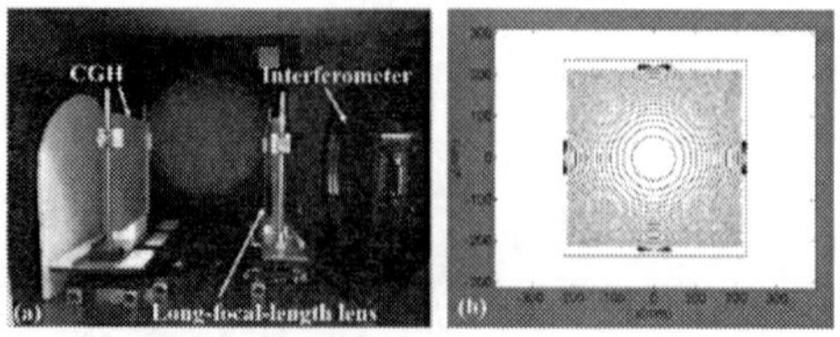

Fig. 2. (a) The experiment setup for testing the long focus lens with CGH; (b) the fringe patterns of multizone CGH.

Figure 3 shows two measurements for 0th and 1st order. We can see that there are ghost fringes near the center of the lens. These ghost fringes are caused by the reflections of lens surface and the diameter are 1.38mm and 49.76mm for the first and second surface, respectively. To calibrate CGH substrate, the 0th order map is transformed to match 1st order map. Figure 4 shows the mapping function between the radius positions of CGH and the measured lens. After CGH substrate calibration, the final transmitted wavefront with CGH substrate calibrated is PV=0.2284λ and RMS=0.0276λ.

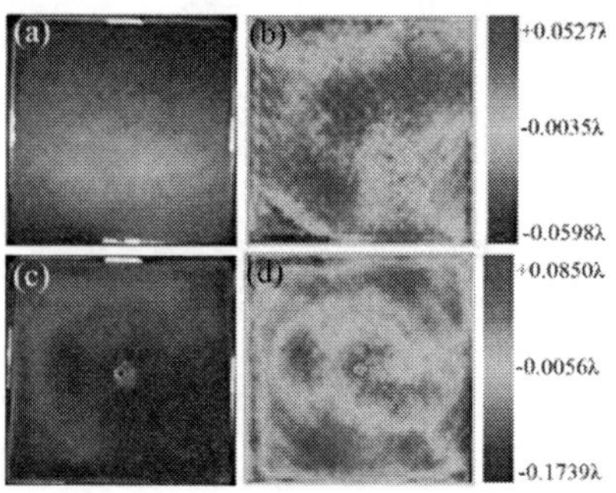

Fig. 3. 0th and 1st order measurements. (a) Interferogram for 0th order; (b) CGH substrate figure map: PV=0.1125λ and RMS=0.0178λ; (c) Interferogram for 1st order; (d) Wavefront map: PV=0.2590λ and RMS=0.0304λ.

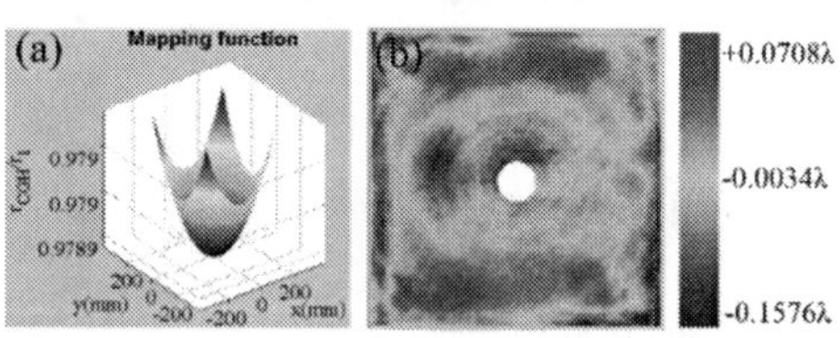

Fig. 4. Data mapping between 0th order and 1st order. (a) Mapping function; (b) Wavefront map with CGH substrate calibrated: PV=0.2284λ and RMS=0.0276λ.

3.2 Error analysis

The CGH test errors includes design error, fabrication errors, alignment errors, and mapping errors, as given in Table 1. Assuming all the errors are un-related and independent, the total wavefront error of this CGH test can be estimated as the root-sum-spuare (RSS) of these errors, which is approximately 0.0091λ RMS. Because the test beam passes through the measured lens twice, the final transmitted wavefront error for long-focal-length lens is half of the RSS value, which is 0.00455λ RMS.

Table 1. Wavefront errors analysis for the CGH test

Source of Errors		Wavefront error (RMS)
Design error of CGH	Design residual	0.0000λ
Fabrication errors of CGH	Pattern distortion error (0.5μm)	0.0051λ
	Etching depth error (5nm)	0.0000λ
	Duty-cycle error (5%)	0.0070λ
	Substrate figure calibration residual (0.0020λ)	0.0020λ
Alignment errors of CGH	*x* tilt (1 fringe)	0.0000λ
	y tilt (1 fringe)	0.0000λ
Fabrication errors of measured lens	Radius of curvature (0.1%)	0.0003λ
	Thickness of lens (0.2mm)	0.0000λ
Alignment errors of measured lens	*x* decenter (0.2mm)	0.0002λ
	y decenter (0.2mm)	0.0002λ
	x tilt (0.04′)	0.0002λ
	y tilt (0.04′)	0.0002λ
	z displacement (1mm)	0.0000λ
Mapping errors	*x* direction (1 pixel)	0.0012λ
	y direction (1 pixel)	0.0012λ
RSS Errors		0.0091λ

Conclusion

A method for measuring the transmitted wavefront of long-focal-length lens with a reflective multizone CGH is proposed. In this method, a reflective CGH with circular lines is employed as an alternative of the convex mirror. This avoids the manufacture of large aperture convex mirror and keeps the advantage of simple test configuration. Four beam projection CGHs are designed to help the pre-alignment of the measured lens, which significantly improves the testing efficiency. Experiments and error analysis exhibit that this multizone CGH approach has a good performance.

References

1. C. Jin, S. Liu, Y. Zhou, X. Xu, C. Wei, J. Shao, *Chin. Opt. Lett.* 12, S21203 (2014).
2. T. G. Parham, T. J. McCarville, and M. A. Johnson, *Optical Fabrication and Testing*, OWD8, (2002).
3. J. C. Wyant, V. B. Bennett, *Appl. Opt.* **11**, 2833, (1972).
4. B. DeBoo, J. Sasian, *Appl. Opt.* **42**, 3903 (2003).

https://doi.org/10.1051/epjconf/201921505001

Effect of Pad Elastic Modulus on The Polishing Induced Plastic Subsurface Damages Distribution of Fused Silica Optics

Xiang He*, *Chao* Cai, and *Ping* Ma

Chengdu Fine Optical Engineering Research Center, Chengdu 610041, China

Abstract. The plastic subsurface damages distribution of fused silica optics polished with different pads are investigated. The elastic interaction model, plastic indentation model and wear relationships are combined together to theoretically characterize the plastic subsurface damages distribution in different polishing processes, which shows consistent results with experiments. It reveals that most of the polishing induced subsurface damages are plastic damages. A few largest polishing particles in the tail end distribution mainly decide the final depth distribution and density of the polishing induced plastic subsurface damages. The larger pad elastic modulus will make the few largest polishing particles bear much larger load and generate larger proportion of observable plastic subsurface damages. Using polishing pad with lower elastic modulus is prominent for restricting the generation of fractures and plastic damages.

1 Introduction

Recent studies suggest that the subsurface damages produced by optical polishing are responsible for initiating the laser induced surface damage of fused silica optics. Several studies focus on predicting the distribution of the polishing induced subsurface damages. Suratwala et al. use the viscoelastic model to characterize the embedding of particles into pad, which could obtain the length of a scratch under different parameters. Liao et al. have developed a model for the interaction of the optics, impurity particle and pad during the polishing process to illustrate the scratch formation process. But these models could not characterize the subsurface damages distribution under different polishing parameters. In addition, the previous researches mainly focus on the polishing induced fractured subsurface damages. Some previous studies also show that the polishing induced subsurface damages contain not only a few fractured damages but also high density plastic damages. But the influence of polishing parameters on the plastic subsurface damages distribution has not been revealed yet.

2 Experiments and results

Five 50 mm diameter, 5 mm thick initial polished fused silica samples were polished with ZrO_2 with different polishing pads (Cloth for sample A, LP13 polyurethane for sample B, LP57 polyurethane for sample C, LP87 polyurethane for sample D, 73# Pitch for sample E). Figure 1 presents the AFM images of all samples after HF etching. Table 1 presents the surface roughness (Rq), the maximum subsurface damage depth (Rmax) and the subsurface damages area percentage for all samples measured by AFM. It can be found that with the increased elastic modulus of polishing pad, the subsurface damages area percentage, the maximum subsurface damage depth and the surface roughness are rapidly increased. The Nomarski images show that most of the polishing induced subsurface damages are plastic scratches, a few fractured scratches are only found in samples D and E.

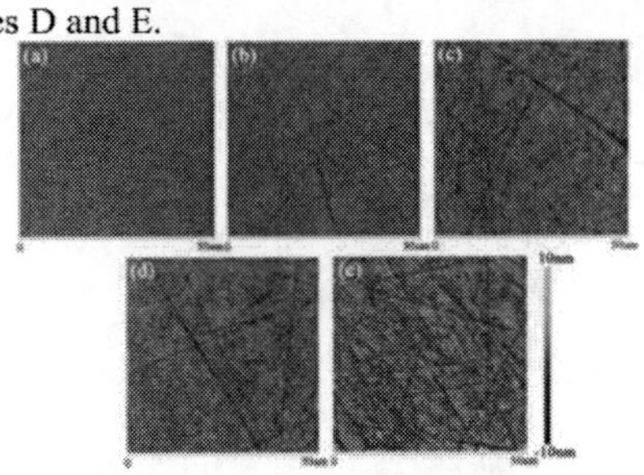

Fig. 1. The AFM images of (a) sample A, (b) sample B, (c) sample C, (d) sample D and (e) sample E with same color scale.

Table 1. The measured subsurface damage distribution.

Sample	The measured surface roughness (Rq nm)	The measured maximum subsurface damage depth (Rmax nm)	The measured subsurface damage area percentage
A	0.53	8.0	0.86%
B	0.74	12.5	2.18%
C	1.87	18.8	8.41%
D	2	22.8	8.92%
E	3.08	55.3	19.27%

3 Discussion

* Corresponding author: rushhe@163.com

If we assume that all the polishing particles are spherical and distributed at the optics–pad interface as a single layer of particles. From Hertzian contact mechanics, the bear load F(r) on each particle with radius r is then given by Eq. (1):

$$F(r) = \frac{16}{9}\frac{E_p}{(1-\upsilon_p^2)}r^{\frac{1}{2}}(2r-g)^{\frac{3}{2}} \quad (1)$$

Where E_p and υ_p are the elastic modulus and Poisson's ratio of the polishing pad, and g is the equilibrium gap at the optics-pad interface. The plastic indentation depth δ_o of the particle into the optics can be calculated by Eq. (2):

$$\delta_o(r) = \frac{8}{9\pi}\frac{E_p}{(1-\upsilon_p^2)\,H_o}r^{-\frac{1}{2}}(2r-g)^{\frac{3}{2}} \quad (2)$$

Where H_o is the hardness of optics. We get the maximum subsurface damage depth (Rmax) from AFM, so we could use Eq. (2) to calculate the equilibrium gap g in the polishing process (as shown in Table 2).

Table 2. The calculated results in all samples.

Sample	equilibrium gap g (nm)	the calculated maximum subsurface damage depth (Rmax nm)	The calculated subsurface damage area percentage	The calculated surface roughness (Rq nm)
A	840	8.0	0.18%	1.15
B	1050	11.35	2.23%	1.82
C	1310	20.18	10.03%	3.23
D	1415	27.47	15.95%	4.56
E	1615	59.35	28.68%	9.96

With the calculated equilibrium gap g, the bear load F(r) and indentation depth δ_o into optics for different particles in different polishing processes can be calculated (as shown in Fig. 2). Then, the instantaneous fraction number ($f_{ins}(\delta_o)$) of scratches with different depth δ_o generated during a certain polishing time could be obtained (as shown in Fig. 3). Considering that with the increased polishing time and material removal (Δ), the previously generated scratches will be shortening continuously and new scratches will be generated. The final fraction number ($f_{fin}(\delta_o)$) of scratches with different depth δ_o can be obtained by Eq. (3):

$$f_{fin}(\delta_o) = \sum_n f_{ins}(\delta_o + n\Delta) \quad (3)$$

If we assume that all the scratches have the same length and included angle. Then we could obtain the final cumulative depth distribution ($S_{cum}(\delta)$) of scratches generated during the polishing process by Eq. (4) (as shown in Fig. 4):

$$S_{cum}(\delta) = \sum_{\delta_o} f_{fin}(\delta_o) \times 2 \times (\delta_o - \delta) \times \tan\frac{\theta}{2} \quad (4)$$

Sum the scratch area deeper than 5nm as the final subsurface damage area percentage (as shown in table 2), which is consistent with the measured results. In Fig. 2, almost all the bear load of polishing particles is smaller than 0.1 N (critical load for fractured deformation). So the model indicates that the micron sized particles will abrasion fused silica optics in plastic deformation region. The few fractured scratches exist in samples D and E are possibly generated by some extremely large interfused impurity particles.

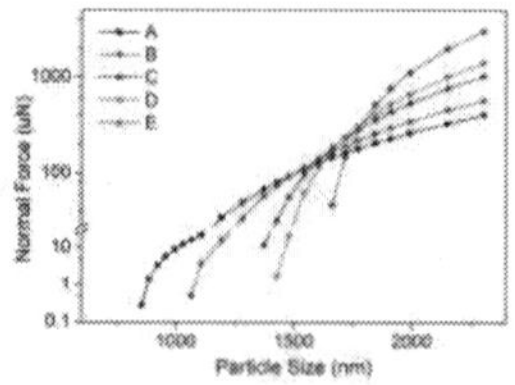

Fig. 2. The calculated bear load of different particles in different polishing processes.

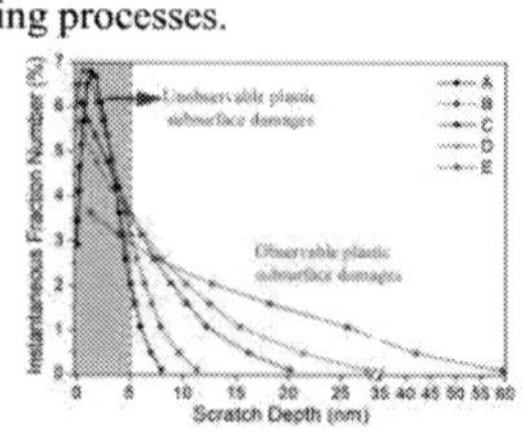

Fig. 3. The calculated instantaneous fraction number of scratches with different depth generated during a certain polishing time.

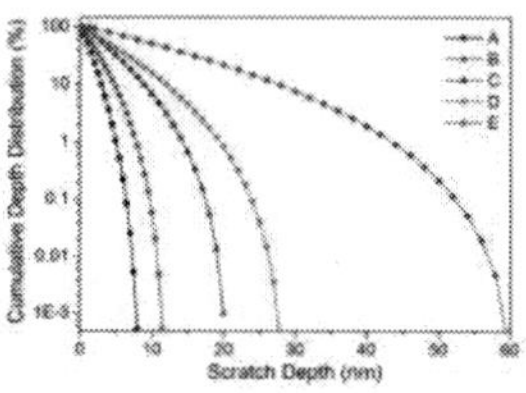

Fig. 4. The calculated final cumulative depth distribution of scratches generated during different polishing process.

In Fig. 2, the bear load and indentation depth increase rapidly with the increased particle size for all samples. The smallest particle size needed to produce observable plastic subsurface damages (deeper than 5 nm) in samples A, B, C, D and E is 1840, 1710, 1663, 1672 and 1736 nm, respectively. It indicates that only 5.4%, 11.6%, 14.3%, 13.6% and 8.9% of the largest polishing particles could initiate observable plastic subsurface damages in samples A, B, C, D and E. So a few largest particles in the tail end of the particle size distribution will mainly determine the final depth distribution and density of the plastic subsurface damages. In Table 2, the smallest diameter of loaded particles increase and the fraction of loaded particles decrease rapidly with the increased pad elastic modulus. It indicates that the few largest particles in the tail end of the particle size distribution will bear much larger load and induce deeper plastic subsurface damages with larger pad elastic modulus. Besides, the fraction of unobservable plastic subsurface damages decreases rapidly with the increased pad elastic modulus. So when polishing pad with larger pad elastic modulus is used, more and more loaded particles participating in polishing will generate observable plastic subsurface damages. Finally, the optics surface will have more and deeper plastic subsurface damages. So using soft polishing pads with lower elastic modulus could substantially decrease the depth and density of polishing induced plastic subsurface damages.

Machine learning robot polishing cell

Max Schneckenburger[1,*], *Luis* Garcia[1], *Rainer* Börret[1]

[1]Aalen University of Applied Science, Centre for Optical Technologies, Aalen, 73430, Germany

Abstract. The quality of optical components such as lenses or mirrors can be described by shape errors and surface roughness. With increasing optic sizes, the stability of the polishing process becomes more and more important. Parameters such as chemical stability of the polishing slurry or tool wear are key elements for a deterministic computer controlled polishing (CCP) process. High sophisticated CCP processes such as magnetorheological finishing (MRF) or the zeeko bonnet polishing process rely on the stability of the relevant process parameters for the prediction of the desired material removal. The aim of this work is to monitor many process-relevant parameters by using sensors attached to the polishing head and to the polishing process. Examples are a rpm and a torque sensor mounted close to the polishing pad, a vibration sensor for the oscillation bearings, as well as a tilt sensor and a force sensor for measuring the polishing pressure. By means of a machine learning system, predictions of tool wear and the related surface quality shall be made. The aim is the detection of the critical influence factors during the polishing process and to have a predictive maintenance system for tool path planning and for tool change intervals.

1 Introduction

With increasing glass optic size, the stability of the polishing process becomes more and more important. The polishing of optical elements depends on several parameters, such as the tilt angle of the tool, the normal force or the friction coefficient, which also influence the material roughness and the material removal rate [1][2][3]. Without controlling the process, the parameters change during the polishing process due to wear. According to the state of the art in the polishing sector, almost no polishing parameters are monitored by sensors. Some parameters such as the normal force or the toolpath velocity are assumed to be constant throughout the polishing process [4]. There will be no stable polishing process on a long period of time without readjusting the polishing parameters. In the past, chemical process conditions assumed to be constant. In the future, additional sensors for measuring chemical values such as temperature, concentration of the polishing slurry and pH value will be installed. These sensors will obtain a better understanding of the chemical polishing process. As a carrier medium deionized water is used with Ceriumoxid, Opaline from Solvey (former Rhodia) with a diameter of 1 µm as polishing grain. The expected pH value, which is influenced by temperature, will be 7.

The main part of the setup for this work (cf. Figure 1) is a polishing head designed with the possibility to use either an eccentric pad, a spinning pad or a polishing wheel; therefore, a comparison of these polishing methods is possible. Currently, the setup with the polishing wheel is finished and first tests were performed, while the setups for the pads are still in construction.

Attached to this polishing head are several sensors that monitor the process data and provide information, such as a voltmeter, that measures the power of the motor, a sensor to measure the torque and rpm of the pad/wheel, as well as sensors to measure vibrations and the normal force. From the data of the voltmeter and the torque sensor/tachometer, the efficiency of the polishing head can be measured. All sensors are connected to a PLC (Programmable Logic Controller), which collects the data. An ABB industrial robot is used to perform the movements of the polishing head.

To test the functionality of the polishing head, BK-7 glass samples are used. For the final tests, flat Schott ZERODUR® (Li_2O-Al_2O_3-SiO_2) samples with a diameter of 90 mm and a height of 30 mm will be polished. Schott ZERODUR® is a common used material for large astronomical mirrors. Furthermore, it is not necessary to monitor the work piece temperature caused by the extremely low coefficient of thermal expansion over a wide temperature range.

For the initial tests with the polishing wheel, the toolpath is programmed to be a straight line forward and backward on the BK-7 glass sample. During the polishing process, the sensor data were monitored and afterwards evaluated using machine learning.

For further testing, the initial surface of the glass will be worked on a lever arm polishing machine to always have the same initial conditions. After a polishing test, the surface can be reset.

* Corresponding author: max.schneckenburger@hs-aalen.de

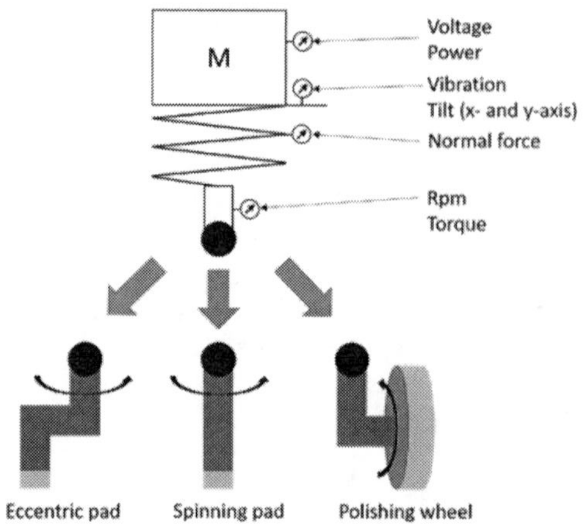

Fig. 1. Principle setup of the polishing head with different sensors and the possibility to use one of three different types of polishing methods (eccentric pad, spinning pad or polishing wheel); the black arrows indicate the movement for each method.

2. Aim of this research

The objective of this investigation is to use the knowledge obtained from previous works and improve the surface quality of glass samples, concerning its surface roughness by using machine learning and different polishing modes.

Initial tests have proven the functionality of the sensors, for example, comparative measurements with a high-precision electronic spirit level have verified the repeatability of the tilt sensor.

Data (cf. Figure 2), monitored by the vibration sensor, have shown a wide range of input values instead of the desired constant values. This effect is caused by the instability of the polishing head, but in that case, a stable polishing process was not possible.
In figure 2, the ripple of all sensor data shows the co- and counterrotating during polishing. The small deflections on the ripple of the sensor data (mainly at the normal force and stroke) come from the uneven surface and unbalance of the polishing wheel, caused by the bond seam of the polishing layer. Another future work is a second version of the polishing head with several improvements. One improvement will be the prevention of the oscillations by increasing the stability. Also, more data from state of the art polishing heads are needed to compare typically normal force, oscillation and efficiency values to the new polishing head. A mechanical stop of the z-axis will improve the safety of the polishing head.

In further tests, synchronspeed polishing should be used just with co- or counterrotating of the polishing head.
For the final tests, a DOE by the Taguchi or factorial method will be done to find the correct polishing parameters and to improve the already existing problems. Also, the DOE data will be used to initial train the machine learning algorithm. By doing more experiments and collecting polishing data, machine learning algorithms will be able to do better predictions of the process. Furthermore, the process will be controlled by changing values for normal force, torque and rpm. In combination with CCP and the inline control of the polishing values, there will be an improvement of the surface quality.

The polishing head will be used for shape correction of optics. With the sensors, the polishing head can be controlled and the parameters of the polishing process can be adjusted. This leads to the possibility of achieving a stable polishing process over a longer period of time.

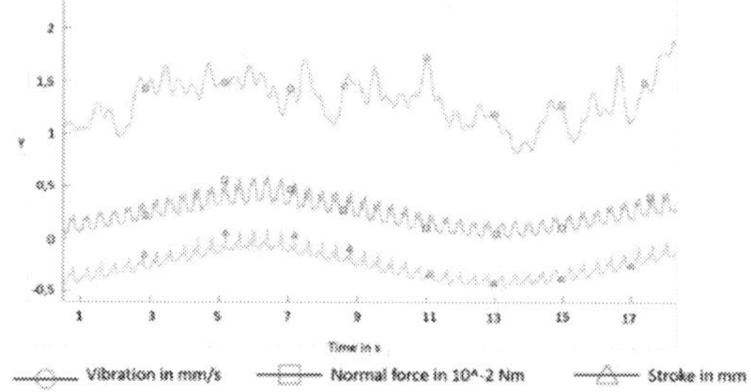

Fig. 2. First sensor data from the polishing head monitored by the PLC (Programmable Logic Controller)

4 Conclusions

First results show the functionality of the polishing head, but also some weak points, which have to be improved. On the basis of these initial tests, further experiments have to be performed to study the influence of different process parameters. Monitoring more data over a longer period of polishing time will allow the use of machine learning algorithms, which leads to a better understanding and control of the polishing process.

References

1. H. Liang, F. Kaufman, R. Sevilla, S. Anjur. „Wear phenomena in chemical mechanical polishing." (1997)
2. J. Luo, D. Dornfeld, „Material Removal Mechanism in Chemical Mechanical Polishing: Theory and Modeling" (2001)
3. J. Bliedtner, G. Gräfe, „Optiktechnologie Grundlagen - Verfahren - Anwendungen – Beispiele", Hanser Verlag (2010), ISBN: 978-3-446-42215-5
4. R. Almeida, R. Börret, D.K. Harrison, A. Desilva, *Influence of different polishing materials in the material removal of steel samples* (Production & Manufacturing Research, 2018)

Steel mould surface improvement by robot fluid jet polishing and robot pad polishing for polymer optic production

Rui Almeida[1,*], *Timon* Ebert[1], *Rainer* Börret[1] and *Mario* Pohl[1]

[1]Aalen University of Applied Sciences, Centre for Optical Technologies, 73430 Aalen, Germany

Abstract. In order to improve the quality of injection-moulded polymer optic parts, it is necessary to improve the quality of the respective steel moulds. For this reason, it is not only necessary to improve the surface roughness of the mould, but also its geometrical shape. The material removal obtained from robot pad polishing is too low. This makes a shape correction after the milling step a very prolonged process. The aim of this work is to use a polishing chain to improve the surface quality of steel samples in terms of shape deviation and surface roughness. This correction polishing chain uses the robot fluid jet polishing for the geometrical shape correction and afterwards the robot pad polishing for the improvement of the surface roughness. Due to the high material removal rates of the fluid jet polishing, it is possible to correct the geometrical shape of steel moulds very fast up to a certain deviation. The pad polishing process improves the surface roughness of the steel samples. A correction of the shape deviation of more than 80% with a RMS of approximately 8 nm was obtained.

1 Introduction

The quality of plastic parts, produced with the plastic injection moulding process, depends on the shape deviation and the surface roughness of the respective mould [1]. This is due to the fact that the produced parts acquire the geometry of the respective mould. Most plastic injection moulding parts use a steel mould tool, which is usually still polished by hand [2]. For this reason the polishing process step accounts for up to 30% of the total costs in the mould manufacturing process. This makes the polishing a very time consuming and expensive process step [3]. The reproducibility of good manual polishing results depends on the workers skill, his form on the day, patience, experience and concentration as mentioned by Klocke [4]. Automatic robotic polishing processes were increasing over the years, replacing manual operations to automated processes as stated by Segreto [5]. Therefore, industry is giving more importance to finishing processes on curved surfaces concerning good surface quality, efficiency and low cost as described by Tian [6]. The surface roughness of steel moulds has already been improved up to 3 nm using the robot as an automatized polishing process [7]. In order to decrease the production time of the final part in the plastic injection moulding process, the steel moulds geometrical shape also needs to be optimized. In previous works [8], it was possible to correct the geometrical shape of flat hardened steel samples from 668.8 nm to 214.2 nm by using the robot pad polishing process. This represents an improvement of 68%. Despite the successful correction and its prediction, this process is very time consuming. The material removal rate per time is very low during the robot pad polishing process. The already known robot fluid jet polishing process is used to reduce the correction time of steel moulds during the polishing phase. The fluid jet polishing process uses water mixed with abrasive grains, which is pumped in a closed circuit, and is jetted with a certain pressure on the surface. Figure 1 shows the fluid jet polishing principle.

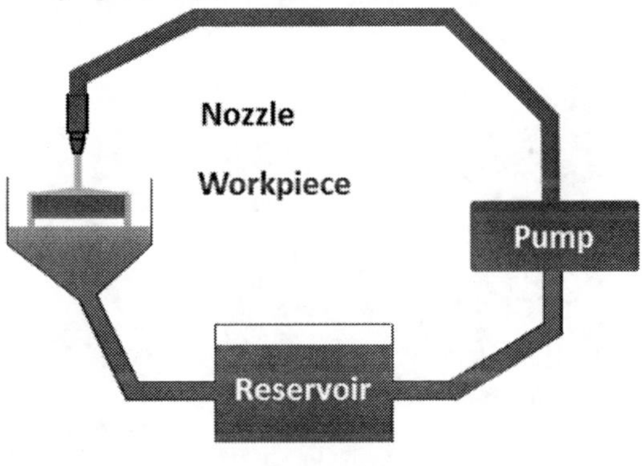

Fig. 1. Principle of the fluid jet polishing process

The material removal on the surface is generated by shear forces that occur upon impact. This polishing process has the advantage of using a very small tool function compared to the robot pad polishing process. Besides that, there is also no wear on the tool compared to the robot pad polishing process. The fluid jet polishing process is used to correct the geometrical shape and the pad polishing process is used to improve the surface roughness of the steel samples, in order to produce a surface with a good quality in terms of shape and roughness.

* Corresponding author: rui.almeida@hs-aalen.de

https://doi.org/10.1051/epjconf/201921505003

2 Aims of this research

The objective of this investigation is to use the knowledge obtained from previous works and improve the surface quality of steel samples, in terms of shape and roughness. The polishing correction technique is applied using the fluid jet polishing process to improve the shape deviation. In this technique, the movement of the robot is manipulated using a feed rate profile. This feed rate profile takes into account the actual shape, the desired shape, the polishing tool function and the desired moving method. The principle of this technique is that on hikes of the surface the movement speed of the robot is decreased in order to create a higher material removal, while on wells the movement speed is increased to create less material removal.

To improve the surface roughness of the steel samples the robot pad polishing process is used. With the current fluid jet polishing process it is only possible to achieve a surface roughness between 200 and 400 nm on a milled surface, for this reason the pad polishing process is used to improve the surface roughness below 10 nm. In order to obtain precise measurements, a Fizeau Interferometer was chosen as optical measuring technique. In order to optically measure a surface a glossy surface is needed. For this reason, the samples were lapped to approximately 10 nm RMS and measured afterwards, afterwards they were fluid jetted and afterwards pad polished. The following figure 2 shows the surface roughness after the fluid jet polishing(A) and after the pad polishing(B). The surface roughness after the fuid jet polishing was between 70 and 100 nm, and after the pad polishing the surface roughness was below 10 nm.

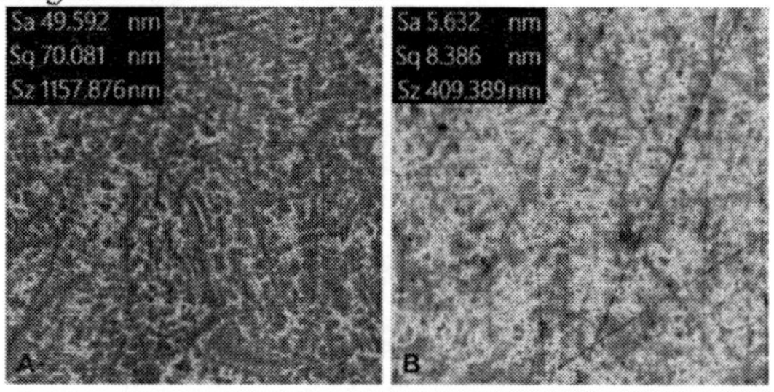

Fig. 2. Surface roughness after fluid jet polishing(A) and after pad polishing(B)

Figure 2 shows the results obtained during the application of the polishing correction technique. The shape deviation before the polishing process was around 963,7 nm and afterwards, the surface was improved up to 115,7 nm.

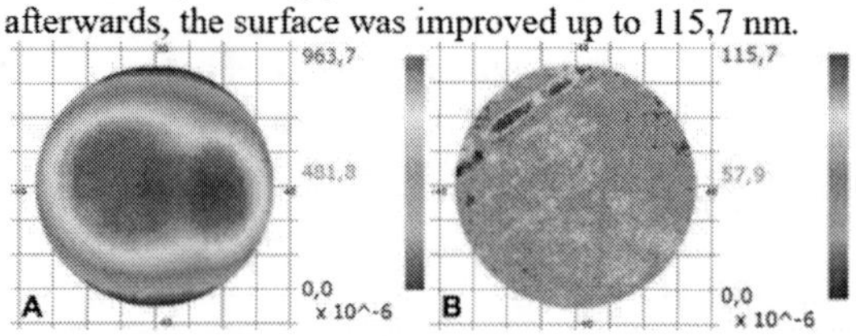

Fig. 3. Shape error before the polishing correction(A) and after polishing correction(B)

3 Conclusions

It was possible to improve the surface quality of steel samples in terms of surface roughness and shape deviation. It was possible to reduce the working/polishing time compared to previous works[8]. Considering the shape deviation, the error of the sample was 963,7 nm in the beginning. After the polishing correction technique it was improved up to 115,7 nm, which is an improvement of more than 85%. This improvement shows the correction capabilities of the fluid jet polishing process as it is possible to create constant and reproducible material removals with a high removal rate per time. Comparing the robot pad polishing process with the robot fluid jet polishing process, there is a material removal factor of 25. This means that the fluid jet polishing is able to produce 25 more material removal in the same amount of time compared to the pad polishing process.

Considering the surface roughness, after the fluid jet polishing the roughness of the sample was between 70 nm up to 100 nm RMS. With the pad polishing process it was possible to improve it up to 5 nm RMS. It was concluded that it is possible to combine the fluid jet polishing process with the pad polishing process for the production of steel moulds. The fluid jet polishing is well suited to improve the shape deviation and the pad polishing to improve the surface roughness. In the end, the steel sample's surface had a RMS of 5 nm and 100 nm shape deviation.

References

1. M. Speich, R. Boerret, *Mould fabrication for polymer optics* (Journal of the European Optical Society: Rapid Publications, 6, 2011)
2. R. Boerret, J. Klingenmaier, U. Berger, A. Frick, *Minimized process chain for polymer optics* (Proceedings of SPIE, San Diego, California, United States, 2008)
3. A. Uddeholm, *Uddeholm Tool Steel for Moulds*, (2016)
4. F. Klocke, O. Dambon, B. Behrens, *Analysis of defect mechanisms in polishing of tool steels* (Production Engineering Research and Development, 2011)
5. T. Segreto, S. Karam, R. Teti, J. Ramsing, *Feature Extraction and Pattern Recognition in Acoustic Emission Monitoring of Robot Assisted Polishing, Procedia* CIRP, 2015)
6. F. Tian, C. LV, Z. Li, G. Liu, Modeling and control of robotic automatic polishing for curved surfaces (CIRP Journal of Manufacturing Science and Technology, 2016)
7. R. Boerret, J. Raab, M. Speich, *Mold production for polymer optics* (Proceedings SPIE, 2014)
8. R. Almeida, R. Boerret, M. Pohl, D. K. Harrison, A. K. M. DeSilva, *Geometrical shape improvement of steel moulds by robot polishing process for polymer optic replication* (Production & Manufacturing Research, 2018)

Study of footprint variations of CCP considering machine kinematics

Guoyu Yu[1*], *Christina* Reynolds[1] *David* Walker[1,3,4] and *Oliver* Fahnle[2]

[1]Laboratory for Ultra Precision Surfaces, University of Huddersfield, G6, TechSpace One, SciTech Daresbury Campus, Keckwick Lane, Daresbury, UK, WA4 4AB
[2] FISBA AG, Rorschacher Str. 268, CH-9012 St. Gallen, Switzerland
[3] University College London, Dpt Physics and Astronomy, Gower St, London, WC1E 6BT, UK
[4] Zeeko Ltd, CoalVille, Leicestershire, LE67 3FW, UK

Abstract. This paper investigates the differentiation between machine's static and dynamic footprint (FP). The results have shown progressing footprint variation related to tool's tilt angle. Continuous tilt angle compensation has been applied to offset this effect.

1 Introduction

Computer controlled polishing (CCP) has been adopted as main optical figuring technology due to its many merits, such as high repeatability, accuracy and efficiency. Today, there are many kinds of CCP machines on the market that based on different material removal mechanism [1, 2, 3]. The consideration of certain technology may base on the processing material, dimension, material removal rate and achievable accuracy.

One important factor to achieve a successful form correction is to maintain consistency of the polishing tool's FPs. It has been observed that Ion beam figuring and fluid jet polishing can reach form accuracies that using fewer iteration steps than those of chemical mechanical polishing (CMP) [4]. The slow converging in form correction are normally caused by the variation of FPs during the polishing process. In our previous publication [5], we have reported second derivative footprint recording (SECondo) of the difference between static and dynamic FP and its compensation. In this paper, we will investigate the trend of SECondo effect with respect to different tool tilt angle. This will provide us with quantitative information to facilitate compensation.

2 Experimental approach

There are many factors that can affect the depth, shape and position of a polishing tool's FP. The dominating first-order factors are head speed, pressure on the part and slurry specific gravity. Factors of second order such as machine runout, temperature, slurry PH have been considered and monitored/controlled. The effect that arises from polishing tool/machine's elasticity and its dynamic behavior has not been studied thoroughly. It has been demonstrated that when the polishing machine is accelerated at very large speed around a small circle, there will be a change of shape between static and dynamic FPs and the amount of shift varies with the tilt angle. This is due to the change of machine dynamics thus the redistribution of tool mass with different tilt angle. In the following experiment, the polishing machine was set to polish circular FPs with continuously-varying tilt angle. The FPs will be analyzed to find an offset value of tilt angle so that the dynamic FPs can be compensated to match the static FP of the same tilt angle.

The static footprint was obtained by keeping tool still whilst the part moves to achieve 3000mm/min relative speed. In contrary, the dynamic footprints were obtained by keeping the part still and the tool moves at 3000mm/min so that dynamic characteristic would be present. The footprint was then measured with interferometer and cross-sections were taken for analysis. Note that in this experiment, the polishing parameters were set to the machine's extreme accelerating allowance so that the dynamic FPs' variation can be observed. To observe a trend in the FPs variation, tilt angle will change continuously. The nominal tilt angle was set to be 20 degrees. When polishing the circular FP, the starting tilt angle was offset by -2 degree and increased to 22 degrees (offset value = 2 degrees) at +Y axis and reduced to 18 degrees at –X axis.

3 Results and analysis

Fig. 1 shows different FPs been plotted together which shows variations of the FPs. The Full Width at Half Maximum (FWHM) of the FPs' center has been calculated. As shown in Fig. 2.

* Corresponding author: author@e-mail.org

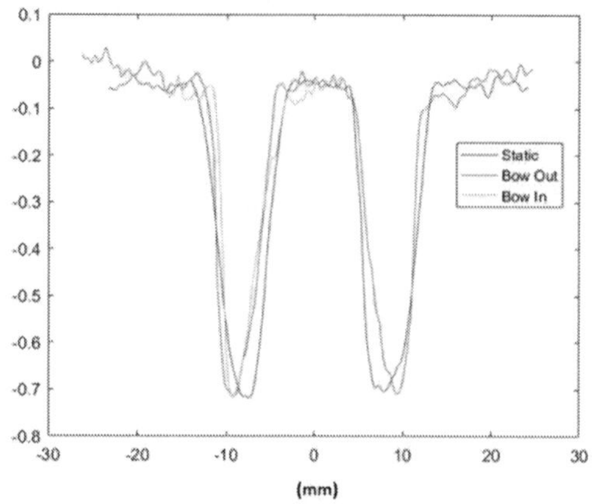

Fig. 1. Dynamic FPs and static FPs showing shape change.

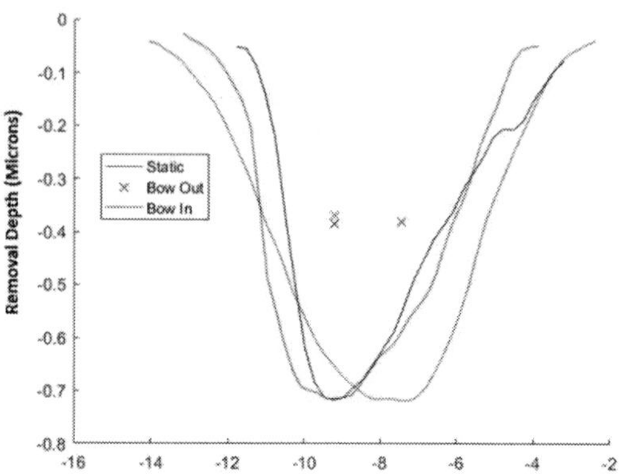

Fig. 2. FWHM analysis of FPs

The X axis cross-section of the circular FP has been used. The tool's tilt angle is 20 degrees. The tool was bowing out so that the tool mass was placed outwards. The coordinates of the centers of the static and dynamic FPs are: X = -7.4651 and -9.2 mm, Y = -0.37954 and -0.36985 µm. The central line of the dynamic and static FPs have been shifted in the X direction by 1.73mm, which will lead to erroneous form correction even if the metrology data has been aligned precisely.

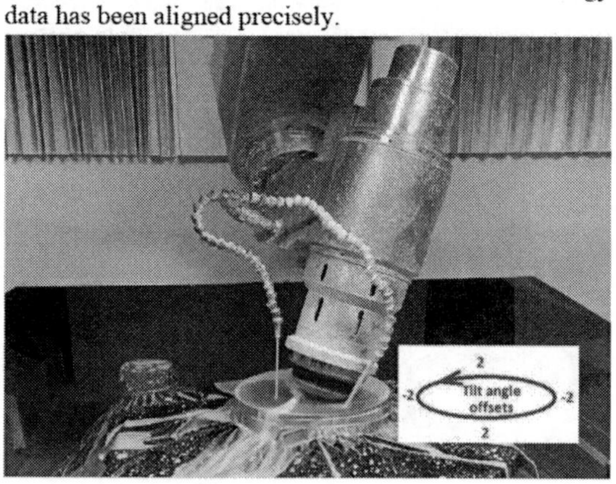

Fig. 3. Experimental setup of tilt angle.

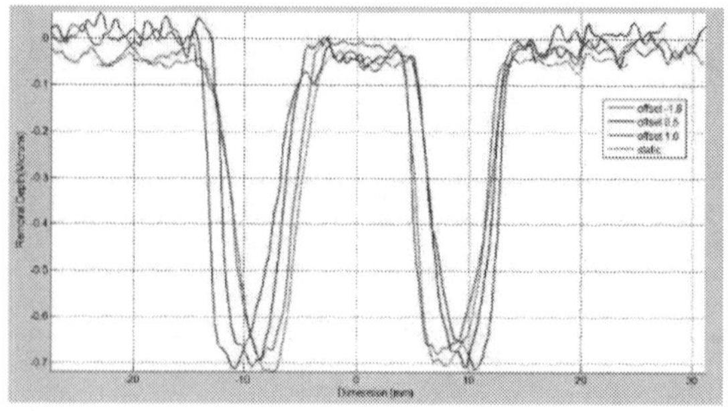

Fig. 4. Dynamic FPs with different tilt angle offsets showing shape change and compensated dynamic FP with static FP (dotted line).

In the experiment, the tilt angle offset setup is shown in Fig.3. From the cross-sections of the FPs (shown in Fig. 4), it can be calculated that the FWHM center of the FPs varies with the change of tilt angle and the best fit of the cross-section is obtained when the tilt angle is 18.2 degrees which are offset the nominal value by -1.8 degrees.

4. Conclusions

We have investigated the shape change of dynamic FPs and found that this can be compensated by tilt angle variation. This is largely because of the change of machine's dynamic distribution. In addition, the combined effect of FP's speed profile and tool run-out have contributed a certain level of compensation.

Reference

1. W. I. Kordonski, S. D. Jacobs, D. Golini, E. Fess, D. Strafford, J. Ruckman, and M. Bechtold, Optical Fabrication and Testing Technical Digest Series 7, 146-149, (1997)
2. D. Walker, R. Freeman, R. Morton, G. McCavana and A. Beaucamp, Optics Express, 14(24), 11787-11795, (2006)
3. T. Franz and T. Hänsel, in Optical Fabrication and Testing, Optical Society of America, Paper OThC7, (2008)
4. O. Fahnle, C. Reynolds, G. Yu and D. Walker, Optical Design and Fabrication 2017, OSA Technical Digest, paper OM4B.1, (2017)
5. O. Fahnle, G. Yu and D. Walker," Third European Seminar on Precision Optics Manufacturing, 1000903, (2016)

EPJ Web of Conferences **215**, 05005 (2019) https://doi.org/10.1051/epjconf/201921505005
EOS Optical Technologies

Surface Figuring of Large Carbon Fiber Reinforced Polymer Antenna Reflector with A Dual-robots Fabrication System

Qiang Xin[1,*], *Haitao* Liu[1], *Jieli* Wu[1], *Lin* Tang[1], *Dailu* Wang[1], and *Yongjian* Wan[1]

[1]Institute of Optics and Electronics, Chinese Academy of Sciences, NO.1 Guangdian Avenue Chengdu, China

Abstract. Carbon Fiber Reinforced Polymer (CFRP) has excellent physical and chemical properties which make it a promising material in making large space borne components, especially in making antenna reflectors and ultra-lightweight space mirrors. These components are usually in large scale to achieve the application requirements. In this research, a dual-robots fabrication system was in-house developed to meet the requirement for figuring a large off-axis parabolic CFRP antenna reflector with the size of 2.4m×4.58m. To make sure that whole surface of the antenna reflector could be covered by the fabrication system, the surface was divided into six regions to accomplish the fabrication. In addition, a special designed tool was utilized to adapt to the curvature variation of the surface. The final surface form accuracies obtained for areas ≤ ϕ 1750mm, ≤ ϕ 2400mm and the whole surface of the antenna reflector were 13.5μm RMS, 23.4μm RMS and 45.8μm RMS, respectively. Feasibility and surface figuring accuracy of the dual-robots system in fabricating large scale components were verified.

1 Introduction

With the development of space observation technology, sizes of the space telescope mirrors and satellite antenna reflectors are becoming bigger to achieve larger observation scope and high resolution. Weight of these space components are also getting larger. Therefore, scientists are facing the challenge of reducing weights of these space components to decrease the launch cost, and maintaining or promoting performances of the components at the meanwhile. Carbon Fiber Reinforced Polymer (CFRP) is a material with excellent physical chemical properties, such as high specific stiffness and low thermal expansion, which proposed an promising solution for making large space borne components with ultra-lightweight, especially in making satellite antenna reflector and ultra-lightweight space mirror[1,2].Size of these components are usually in scale of meters or even tens of meters which proposes great challenge in figuring their surfaces. In this research, a flexible dual-robots fabrication system was in-house developed to meet the requirements of figuring a large off-axis parabolic CFRP antenna reflector. The robots could work collaboratively in machining the reflector surface which could effectively raise the surface figuring efficiency and lower the time and economic costs.

2 The CFRP antenna reflector

Dimension of the off-axis parabolic CFRP antenna reflector to be figured is 2400mm×4580mm as shown in Fig. 1. Surface of the reflector has a radius of curvature of 7992mm. The off-axis distance is 3898mm. Typical sandwich structure is utilized by the antenna whose skin and core are both made of CFRP. Study shows that this design has better performance for space borne antenna reflector in resisting the thermal deformation [1].

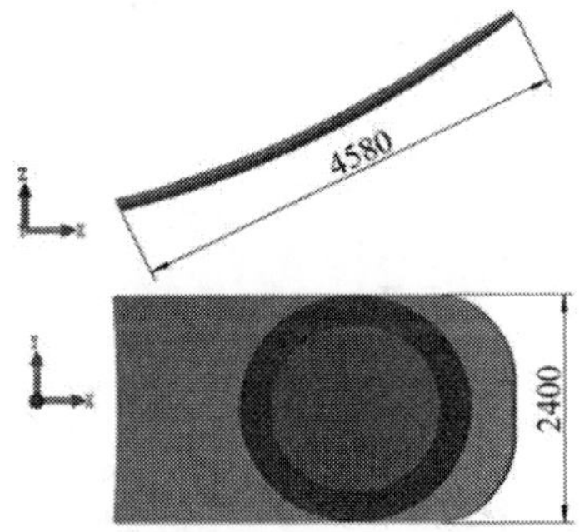

Fig. 1. CFRP antenna reflector

Root Mean Square error(RMS) of the surface form accuracy requirements are varied in different areas of the antenna reflector. Accuracies for areas ≤ ϕ 1750mm (as the red region shown in Fig.1), ≤ ϕ 2400mm (as the red and blue regions shown in Fig.1) and the whole surface are 15μm RMS, 30μm RMS and 90μm RMS, respectively.

3 Dual-robots fabrication system and surface figuring method

Arm length of the robot used in this research is 2.8m. However, the effective workspace of the robot is less than

* Corresponding author: xinqiang@ioe.ac.cn

 https://doi.org/10.1051/epjconf/201921505005

2.8m due to its position and posture when processing the antenna reflector. Moreover, the stiffness and the stability of the robot are decreased when robot working with its arm fully extended. Therefore, it is necessary to use two robots to fabricating the antenna cooperatively under its safe workspace with high stiffness and stability.

In this research, two robots were employed to achieve the surface figuring requirements. The robots were mounted on the rail guides besides the antenna as shown in Fig. 2. Each robot figured half surface of the antenna. They were positioned at different positions on the rail guides to fulfil the surface fabrication. Their positions were computed to assure that the whole surface of the antenna could be covered by the workspace of the robots. Fig. 2 presents the positions RA1, RA2 and RA3 for Robot No.1, and RB1, RB2 and RB3 for Robot No.2. Relations were then determined by accurately calibrating the antenna under coordinate of robots at different positions before the surface fabrication.

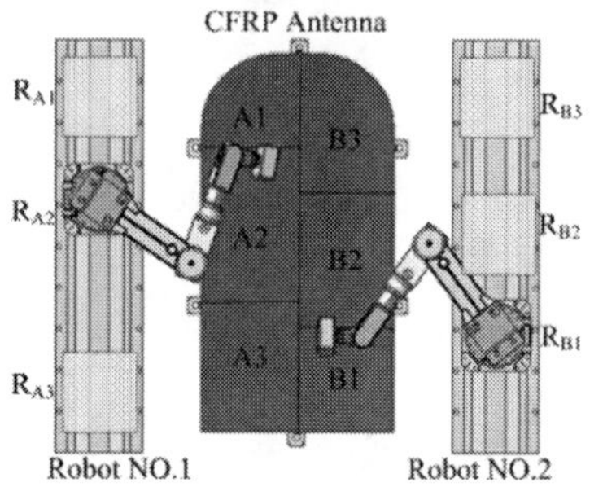

Fig. 2. Schematic of dual-robots fabrication system

A flexible lapping tool was employed to figure the surface of the antenna as shown in Fig. 3. The lapping tool could quickly adapt to the rapid change of surface curvature of the component. The tool moved in a spinning and orbital motion as the traditional one used in computer controlled optical surfacing. A classical meander raster-scanning tool path was adopted in figuring the antenna surface to obtain the required surface form accuracy. In order to lower the traces being produced in between the adjacent machining regions, overlap of the lapping tool in boundaries of the adjacent regions was deliberately considered in planning the surface figuring trajectory.

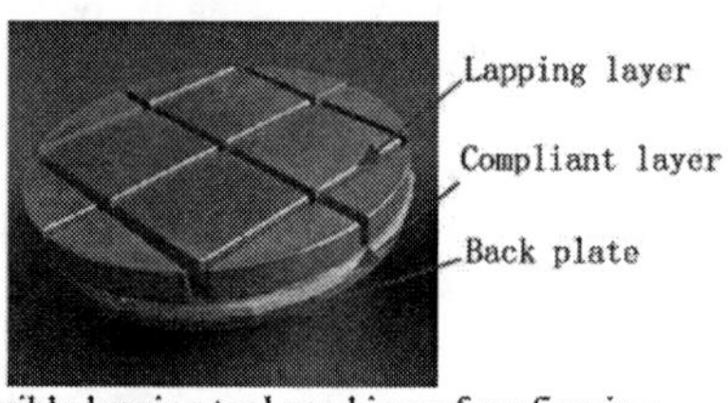

Fig. 3. Flexible lapping tool used in surface figuring

Because the surface form accuracy requirements were higher at the central areas (≤ ϕ 1750mm and ≤ ϕ 2400) than the rest one. These areas were extracted and handled specially to reach their accuracy targets.

4 Results and discussion

Surface form accuracy of the fabricated surface was measured by Radian laser tracker produced by Automatic Precision Inc. Before figuring, the surface form accuracies for areas ≤ ϕ 1750mm, ≤ ϕ 2400mm and the whole surface were 42.3μm RMS, 51.7μm RMS and 56.9μm RMS, respectively. After figured by using the dual-robots system, the accuracies for areas ≤ ϕ 1750mm, ≤ ϕ 2400mm and the whole surface were improved to 13.5μm RMS, 23.4μm RMS and 45.8μm RMS, respectively. Fig. 4 presents the surface form accuracies before and after the fabrication.

Because surface form accuracy at area≤ ϕ 1750mm was higher, more material was removed to achieve its accuracy. Therefore, there is an obvious boundary between area≤ ϕ 1750mm and areas≤ ϕ 2400mm as shown in Fig. 4 b). To eliminate this phenomenon, an effective solution is to figure the central area firstly and reach its surface form accuracy, and then figure the rest surface with comparative material removal rate.

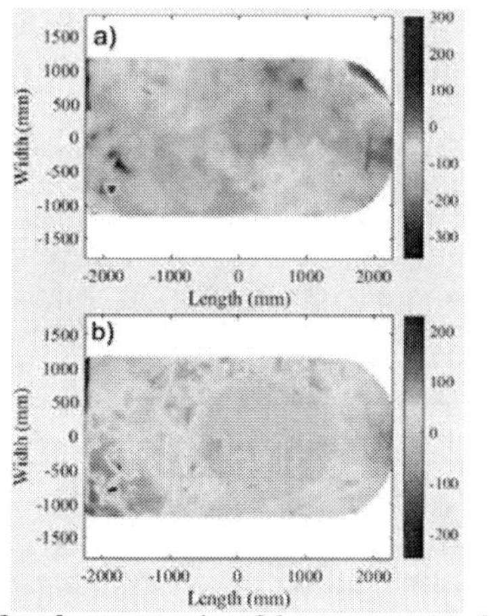

Fig. 4. Surface form accuracies of the antenna, a) before surface figuring and b) after surface figuring

5 Summary

A dual-robots fabrication system was in-house developed to meet the surface figuring requirements for a large off-axis parabolic CFRP antenna reflector with the size of 2.4m×4.58m. The final surface form accuracies figured by using this fabrication system reached the accuracy targets. Flexibility and figuring accuracy of the fabrication system in processing large components were verified.

References

1. X. Zhou, X. Zhou, D. Du, X. Kong, G. Qin, Spacecraft Eng. **27**, 83 (2018)
2. S. Utsunomiya, T. Kamiya, R. Shimizu, *Proc. of SPIE,* **8837**, 88370P (2013)

Development and Application of MRF Based on Robot Arm

Zhang Xuejun[1,2] *, Li Longxiang[1,2**] , Xue Donglin[1,2] , Song Chi[1,2], , Ai Bo[1,2]

[1]Key Laboratory of Optical System Advanced Manufacturing Technology, Chinese Academy of Sciences, Changchun 130033, China
[2] Changchun Institute of Optics, Fine Mechanics and Physics, Chinese Academy of Sciences, Changchun 130033, China

Abstract. Magnetorheological Finishing (MRF) is widely regarded as an effective technique to polish and figure aspheric optics. MRF based on Robot Arm is developed by us. This new machine is more flexible, efficient, cost-effective and smaller space-usage in optical shop, compared with the traditional MRF machines. The components of MRF based on Robot Arm are introduced firstly. Position-attitude control and polishing tool path are also studied in this paper. The experiments and application of MRF based on Robot Arm demonstrate the effectiveness and validity of MRF based on Robot Arm in optical fabrication.

1 Introduction

Magnetorheological Finishing (MRF) is a typical, deterministic sub-aperture computer-controlled optical-surfacing technology [1] and widely regarded as an effective means of processing aspheric optical components due to its high deterministic tool influence function(TIF) and excellent surface roughness quality[2-5]. MRF is originally invented by Dr. William Kordonski and his colleagues in Belarus and commercialized by QED Technologies in USA [6]. In those commercialized MRF machines, such as Q22-950 and Q22-1200[7], a module of MRF is integrated into traditional machine which is called 'Box-machine' of MRF in this paper. The Box-machine of MRF is characterized by high position-accuracy and high work-stability. They has proved the capability of MRF by high-precision and fast polishing aspherical optics. However, such Box-machine has some weakness in optical shop. For example, Box-machines has an expensive price and high equipment cost. A large space is necessary for Box-machines. Moreover, the machine motion performance including velocity and acceleration, is limited when the size of Box-machine become large.

In order to solve these problems mentioned above in optical shop, we developed a new type of MRF machine based on Robot Arm. A specially designed MRF module was integrated into the robot arm who adopts a 6-axis series structure. This type of MRF machine has high motion performance and six degrees of freedom, which is flexible enough to manufacture aspherical or free-from optics. It also has a smaller space occupation and lower equipment cost compared with Box-machine. These advantages make the MRF based on Robot Arm a high cost-effective polishing equipment.

2 Components of MRF based on Robot Arm

The working wheel of the MRF module is Φ360mm and Φ 160mm with a permanent magnet[8]. The MRF module is integrated into the Robot Arm as shown in Fig. 1(a) and (b). The layout of Magnetorheological Fluid cycle system was carried out reasonably to ensure the stability of the whole MRF system on Robot Arm.

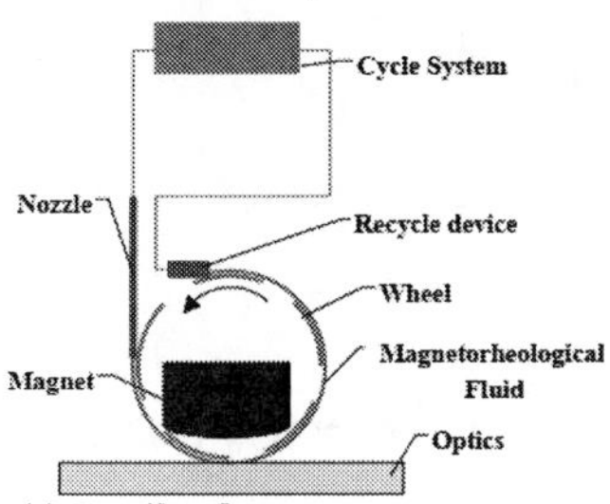

(a) Details of components

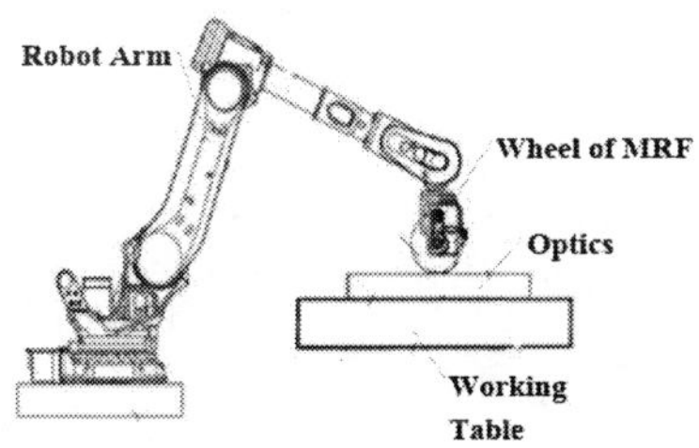

(b) The removal function of MRF

Fig. 1 The MRF based on Robot Arm

3 Position-attitude control of MRF based on Robot Arm

One of the position-attitude control principles for the MRF process requires that the normal direction of the outer surface of the MRF working wheel is aligned with the normal direction of the local aspherical surface[9]. Fig. explains the ZYX Euler angle of the Robot Arm. The definition of rotation transformation is shown in Equation (1).

* Corresponding author: zxj@ciomp.ac.cn; **lilx@ciomp.ac.cn

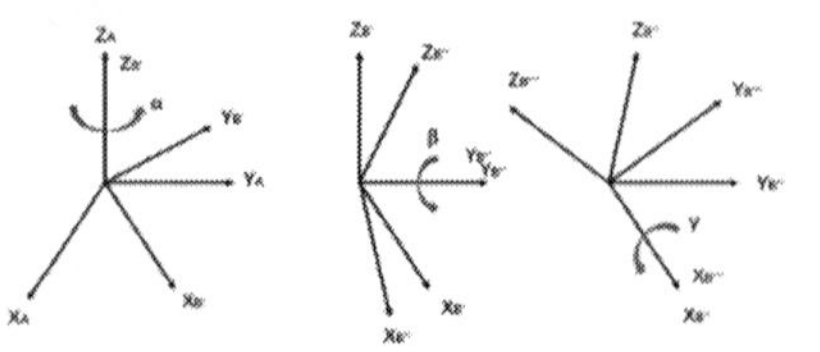

Fig. 2 ZYX Euler angle

$$
{}^{A}_{B}R_{Z'Y'X'} = R_Z(\alpha)R_Y(\beta)R_X(\gamma) = \begin{bmatrix} c\alpha & -s\alpha & 0 \\ s\alpha & c\alpha & 0 \\ 0 & 0 & 1 \end{bmatrix} \begin{bmatrix} c\beta & 0 & s\beta \\ 0 & 1 & 0 \\ -s\beta & 0 & c\beta \end{bmatrix} \begin{bmatrix} 1 & 0 & 0 \\ 0 & c\gamma & -s\gamma \\ 0 & s\gamma & c\gamma \end{bmatrix} \quad (1)
$$

This method of transforming angle of MRF based on Robot Arm, is meaningful in theoretical calculation and initial simulation. But sometimes it is not enough to use the system to conduct practical optical processing. It is often needed to introduce quaternion or some parameter restrictions on the rotation direction of a specific axis to achieve the unique solution of the robot, so as to achieve the purpose of making the actual trajectory of the system consistent with our requirements. The transformation equation from Euler angle to quaternion is shown in Equation

$$
\begin{bmatrix} Q1 \\ Q2 \\ Q3 \\ Q4 \end{bmatrix} = \begin{bmatrix} c\left(\frac{\alpha}{2}\right)c\left(\frac{\beta}{2}\right)c\left(\frac{\gamma}{2}\right) + s\left(\frac{\alpha}{2}\right)s\left(\frac{\beta}{2}\right)s\left(\frac{\gamma}{2}\right) \\ s\left(\frac{\alpha}{2}\right)c\left(\frac{\beta}{2}\right)c\left(\frac{\gamma}{2}\right) - c\left(\frac{\alpha}{2}\right)s\left(\frac{\beta}{2}\right)s\left(\frac{\gamma}{2}\right) \\ c\left(\frac{\alpha}{2}\right)s\left(\frac{\beta}{2}\right)c\left(\frac{\gamma}{2}\right) + s\left(\frac{\alpha}{2}\right)c\left(\frac{\beta}{2}\right)s\left(\frac{\gamma}{2}\right) \\ c\left(\frac{\alpha}{2}\right)c\left(\frac{\beta}{2}\right)s\left(\frac{\gamma}{2}\right) - s\left(\frac{\alpha}{2}\right)s\left(\frac{\beta}{2}\right)c\left(\frac{\gamma}{2}\right) \end{bmatrix} \quad (2)
$$

3 Application of MRF based on Robot Arm

A 1000 mm×470 mm off-axis aspherical mirror is also processed by MRF based on Robot Arm. The material of the mirror is also SiC with a layer of Si on the surface. The initial surface accuracy is 45.76 nm rms as shown in Fig. (a). Considering the risk for this project-used mirror, a high-precision residual surface error is not pursued within only one polish-test iteration, therefore the total time of one iteration is set to 7 hours. After 7 hours effectively polishing as shown in Fig. (b), the rms of the surface error map of the mirror is converged to 22.78 nm as shown in Fig. (c). Although the convergence rate is about 50.2%, the determinacy of MRF based Robot Arm is also very high considering the limit of total polishing time in one cycle. This proves the MRF based Robot Arm is also effective and valid for aspherical optics.

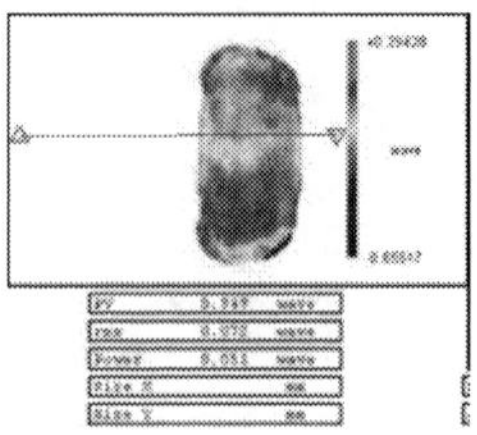

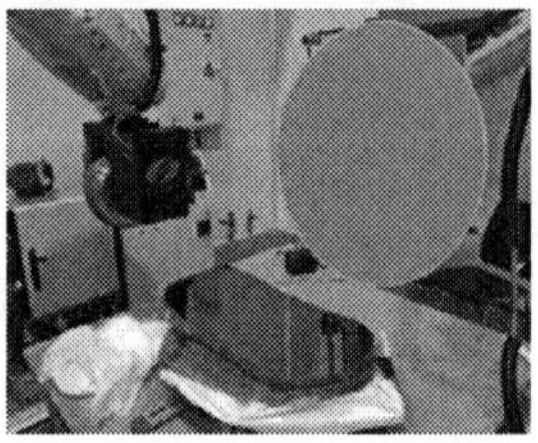

(a) I nitial surface error map (b) Actual process by MRF

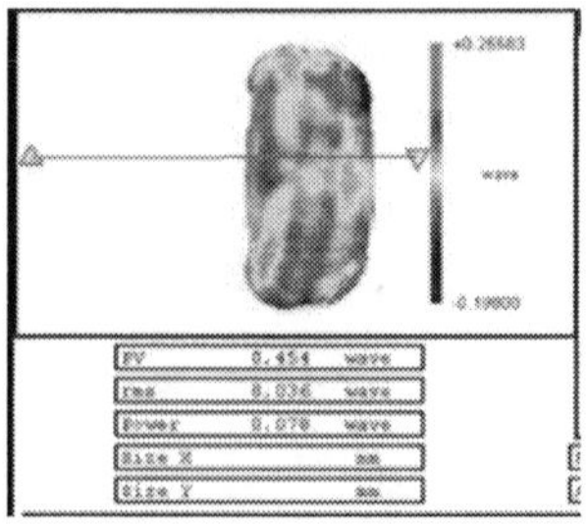

(c) The residual surface error map

Fig. 3 Off-axis aspherical mirror is processed 7 hours

4 Conclusion

Developments and application of MRF based on Robot Arm are introduced in this paper. The application in real optics fabrication proves that this new MRF based on Robot Arm is effective and valid for high-precision plane and curved optical surface. Compared to traditional MRF machine, the MRF based on Robot Arm developed by us is more flexible, efficient, cost-effective and smaller space-usage in optical shop. More work about MRF and robot arm will be done in the future.

References

[1] Li L, Xue D, Deng W, et al. Positive dwell time algorithm with minimum equal extra material removal in deterministic optical surfacing technology.[J]. Applied Optics, 2017, 56(32):9098.

[2] Kordonski W I , Jacobs S D . Magnetorheological Finishing [J]. International Journal of Modern Physics B, 1996, 10(23n24):2837-2848.

[3] Golini D, Kordonski W I, Dumas P. Magnetorheological finishing (MRF) in commercial precision optics manufacturing[J]. Proceedings of SPIE - The International Society for Optical Engineering, 1999, 3782:80-91.

[4] Golini D, Kordonski W I, Dumas P. Precision optics fabrication using magnetorheological finishing[J]. Critical Reviews, 2017, 67:251-274.

[5] Yang C, Peng X, Hu H, et al. Stability Control of Magnetorheological Finishing Removal Function Under Continuous Process Status[J]. Nanotechnology & Precision Engineering, 2016.

[6] Harris, Daniel C. History of Magnetorheological Finishing [J]. Proceedings of SPIE - The International Society for Optical Engineering, 2011, 8016(4):561-566.

[7] https://qedmrf.com/en/mrfpolishing/mrf-products/q22-950

[8] Li L, Zheng L, Deng W, et al. Optimized dwell time algorithm in magnetorheological finishing[J]. International Journal of Advanced Manufacturing Technology, 2015, 81(5-8):833-841.

Figuring of optical aluminium devices by reactive ion beam etching

Jens Bauer[1,*], *Melanie* Ulitschka[1], *Frank* Frost[1], and *Thomas* Arnold[1,2]

[1]Leibniz Institute of Surface Engineering (IOM), Permoserstrasse 15, D-04318 Leipzig, Germany
[2]TU Dresden, Institute of Manufacturing Science and Engineering, George-Bähr-Strasse 3c, D-01062 Dresden, Germany

Abstract. Ion beam figuring (IBF) is an established method in high-end surface manufacturing. However, the direct processing of desired materials as standard Al alloys (e.g. Al6061) fails, since the surface roughness increases drastically as a result of inhomogeneous etching due to structural, crystallographic and chemical irregularities inside the material matrix. As an alternative figuring technology reactive ion-beam etching (RIBE) is a promising route. RIBE provides the direct machining of Al alloys while preserving the surface roughness almost in its initial state. The RIBE process with nitrogen gas is focused more detailed in this study.

1 Introduction

Metal optics are of increasing interest, since refractive optical elements reach their physical limitations within the technological progress in EUV/XUV lithography, x-ray and synchrotron optics. The upcoming needs of most flexible shape adjustment and increasing surface quality demand improved deterministic tools in ultra-precision surface machining. RIBE is a promissing technique for figure error correction of Al mirror devices. In addition to conventional IBF the process gas contains chemically active species as oxygen [1] or nitrogen [2] changing the etch mechanism from pure sputtering to a chemical-assisted sputter etch process. The direct ion-beam machining process has been optimized for figure error correction of standard Al alloy surfaces with focus on superior optical surface quality.

2 Experimental

Planar single-point diamond turned (SPDT) Al samples (supplied from IOF Jena) were made from the RSA alloys Al6061 and Al905 (RSP technology).

For figure error correction low-energy, narrow ion beams (≤ 1.5 keV) are used in deterministic operation via a dwell-time approach. Aperture-free ion beams with a Gaussian shape and a full-width at half maximum (FWHM) of about 5-7 mm and sub-aperture ion beam tools with FWHM ≥ 0.6 mm are available [3]. Latter allows figure error correction from the low-spatial frequency to the mid-spatial frequency range $< 0.5\,\text{mm}^{-1}$. By use of a six-axes CNC motion setup for sample movement, also deep and arbitrarily shaped devices can be corrected [4].

The surface morphology is analyzed by optical microscopy, white light interferometry (WLI), and atomic force microscopy (AFM). Power spectral density (PSD) functions are calculated for roughness analysis.

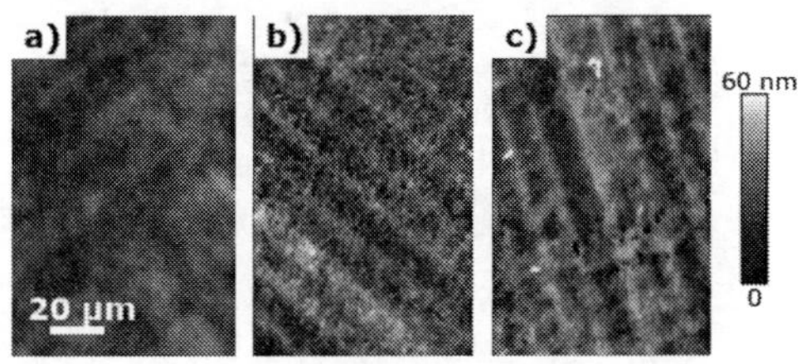

Figure 1. Surface topographies by WLI: a) an untreated Al905 surface after SPDT with a rms roughness of R_q = 10.9 nm, b) argon ion treatment (1.5 keV, etching depth: 330 nm) results in considerable roughening with R_q = 21.0 nm, and c) nitrogen ion treatment (1.5 keV, etching depth: 1 μm) with an almost preserved initial topography at R_q = 12.2 nm.

3 Results and discussion

RIBE processing with oxygen and nitrogen as reactive gas species allows the direct figuring of Al surfaces [1, 2]. In contrast to the classical RIBE scheme, no volatile process products are generated, but the machined surface is determinately chemically modified. In particular, an in situ forming surface layer acts as an etch front moderating the inhomogeneous structural conditions of the Al matrix. Etching occurs as physical sputter erosion within this surface layer exclusively. A very stable and non-hazardous etch mechanism is obtained. As a remarkable technological achievement, the surface roughness is preserved almost in its initial state during RIBE processing while conventional IBF with argon gas results in a considerable surface roughening (Fig. 1). Thus, the RIBE technique permits ultra-precision figure error correction of high-quality optical Al surfaces.

The RIBE process conditions have been investigated in more detail with a special focus on the preservation of the initial topography (Fig. 2). The ion implantation depth depends on the ion energy. Thus, an effect on the ni-

*e-mail: jens.bauer@iom-leipzig.de

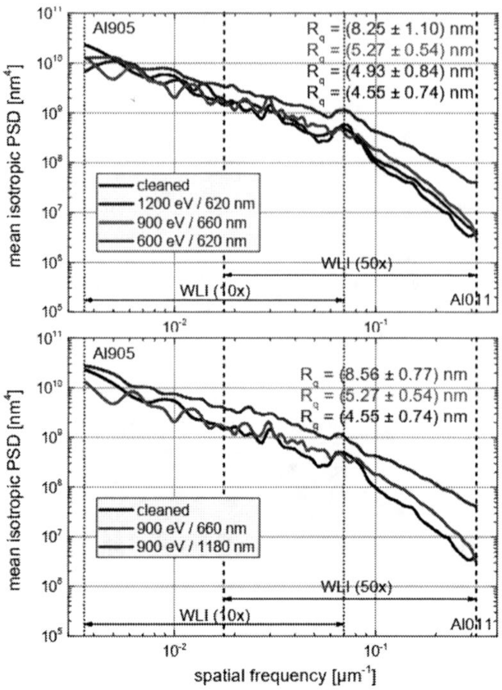

Figure 2. PSD functions before and after RIBE processing with nitrogen gas. Surface degradation increases at lower ion energies (upper) and for increased process time and thus depth (lower).

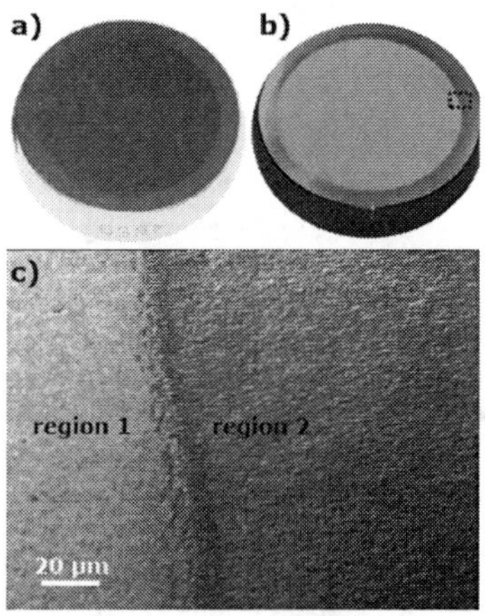

Figure 3. a) Aluminium sample after nitrogen ion beam treatment (1.5 keV) applying a graphite contact mask with 38 mm opening. b) The nitride layer was removed by wet-chemical processing with a 1% KOH solution. c) Microscopy image of the marked area in b). Region 1 defines the sample area, which was first ion beam treated and subsequently wet-chemically etched. Region 2 indicates the area, which was masked during ion beam treatment and thus only wet-chemically etched.

tride layer formation can be expected. At low ion energies <900 eV some surface degradation was observed already at ~600 nm etch depth. For ion energies >900 eV the initial topography is almost preserved up to 1 μm etch depth.

However, the surface nitride lowers the reflection properties of the Al surface. Thus, the nitride removal by wet-chemical treatment has been focused. AlN is highly resistant against most chemical etchants including HCl. Considerable AlN etching performance is provided only by KOH-based etch solutions and to minor extent by NaOH-based etchants. A sequence of treatment in 1% aqueous KOH solution and subsequent rinsing in DI-water has been applied (Fig. 3a/b). The etching rate is 4.4 Å/s. In Fig. 3c the border area of nitride layer (region 1) and clean aluminium surface (region 2) is shown after wet-chemical etching. Since the etch stop has been chosen very accurately to the nitride layer thickness the initially nitrided region 1 exhibits a very smooth surface free of any nitride residuals. In contrast, the clean aluminium degraded considerably as a result of strong etching. Hence, a precise etch stopping is necessary for successful nitride removal.

In summary, for direct surface figuring of optical Al surfaces the RIBE process conditions with nitrogen gas are addressed. At ion energies >900 eV the topography is preserved up to 1 μm etch depth. With decreasing ion energy and increasing process time surface degradation issues need to be accepted. However, the formation of a homogeneous surface nitride layer is observed for any process condition. Since the optical surface properties are diminished by the nitride, a wet-chemical etching approach for the removal of the nitride layer is successfully applied.

T. Liebeskind is thanked for wet-chemical sample treatment. Financial support given by the German Federal Ministry of Education and Research (BMBF) within InnoProfile Transfer 03IPT706X is gratefully acknowledged.

References

[1] J. Bauer, F. Frost, T. Arnold, J. Phys. D. Appl. Phys. **50**, 085101 (2017)

[2] J. Bauer, M. Ulitschka, F. Frost, T. Arnold, *Figure error correction of aluminium mirrors by deterministic reactive ion-beam machining*, in *EOS Opt. Technol. / 5th Conf. Manuf. Opt. Syst.* (European Optical Society, Munich / Germany, 2017)

[3] J. Bauer, M. Ulitschka, F. Frost, T. Arnold, L. Alber, M. Sondermann, G. Leuchs, *Ultra-precision surface figuring of optical aluminium devices*, in *OSA Adv. Photonics Congr. / Nov. Opt. Mater. Appl. Top. Meet.* (Optical Society of America, Zurich / Switzerland, 2018)

[4] J. Bauer, M. Ulitschka, F. Pietag, T. Arnold, J. Astron. Telesc. Instrum. Syst. **4**, 046003 (2018)

Testing and characterization of challenging optics and optical systems with Shack Hartmann wavefront sensors

Julie Siv[1], * *Rafael* Mayer, *Guillaume* Beaugrand, *Guillaume* Tison, *Rémy* Juvénal, *Guillaume* Dovillaire

[1]Imagine Optic, 18 rue Charles de Gaulle 91400 Orsay, France

Abstract. The Shack-Hartman wavefront sensor is a common metrology tool in the field of laser, adaptive optics and astronomy. However, this technique is still scarcely used in optics and optical system metrology. With the development of manufacturing techniques and the increasing need for optical characterization in the industry, the Shack-Hartmann wavefront sensor emerges as an efficient complementary tool to the well-established Fizeau interferometry for optical system metrology. Moreover, the raise of smart vehicles equipped with optical sensors and augmented reality, the optical characterization of glass and transparent flat materials becomes an issue that can be addressed with Shack-Hartmann sensors. Aberration measurements of challenging optics will be presented such as optical filters, thin flat optics, aspheric lenses and large optical assemblies.

1 Comparison between a Shack-Hartmann and a Fizeau interferometer

1.1 Shack-Hartmann Principle

In a Shack-Hartmann (SH) [1] wavefront sensor a microlens array is placed in front of a detector and each illuminated microlens is focusing on a dedicated part of the detector. The position of each focal spot is compared to a reference position, determined during calibration, gives information about the local slope of the wavefront which is used to reconstruct the aberration map of an optical system.

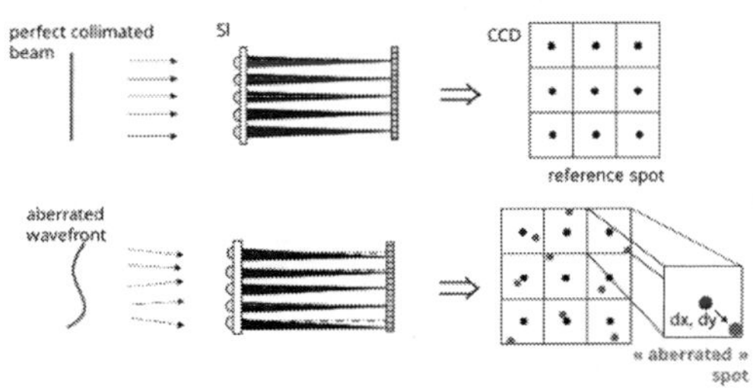

Fig. 1. Principle of Shack-Hartmann wavefront sensing, showing a perfect (top) and aberrated (bottom) wavefront.

1.2 Comparison of a Shack-Hartmann wavefront sensor and a Fizeau interferometer

We performed a comparison between a Fizeau interferometer and a SH wavefront sensor by characterizing a large flat substrate under the same conditions. To be able to do so, we used HASO RFLEX Large aperture, a metrology tool based on SH wavefront detection with an integrated light source and a collimator, allowing to make a measurement on a pupil of 100 mm in diameter.

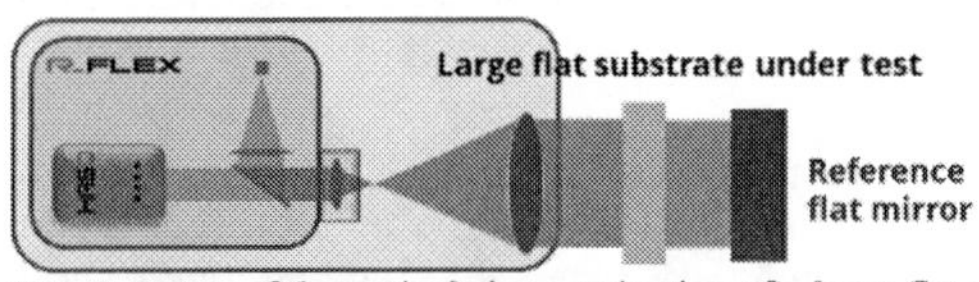

Fig. 2. Setup of the optical characterization of a large flat glass (yellow slab) with a Shack Hartmann wavefront sensor

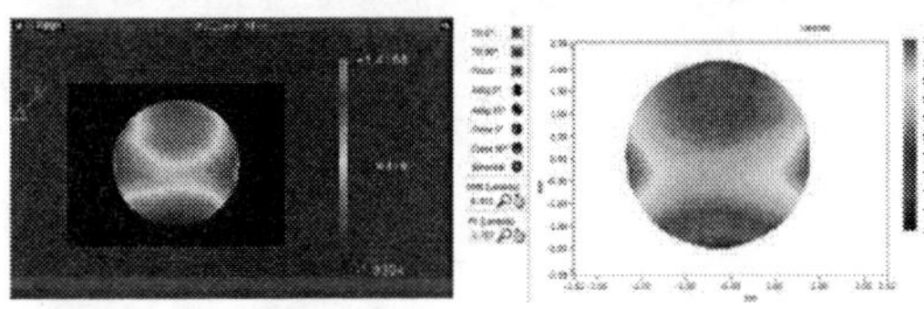

Fig. 3. (Left) measurement made with a Fizeau interferometer (reference model), the computed wavefront error is 2.74 λ PV. (Right) measurement made with a SH wavefront sensor, the computed wavefront error is 2.75 λ PV.

This result shows that there is an excellent agreement between the two measurements.

2 Optics and optical system characterization and testing with SH wavefront sensors

In this part we will show examples of aberration map of a set of optical components and systems that typically are difficult to measure with common interferometric measurements. These measurements were made with wavefront sensors from Imagine Optic and shows that SH wavefront sensors are good complementary tools for optics metrology.

* Corresponding author: jsiv@imagine-optic.com

2.1 Optical filters

The SH wavefront sensor is achromatic. Thus, it is possible to characterize aberrations of an optical component according to the wavelength without changing the metrology tool.
We measured aberrations induced by a large optical filter in transmission at 632 nm and in reflection at 1064 nm. We managed to measure the wedge and the curvature of the optics, which were the dominant aberrations. By filtering these two parameters, we have been able to measure the residual aberrations.

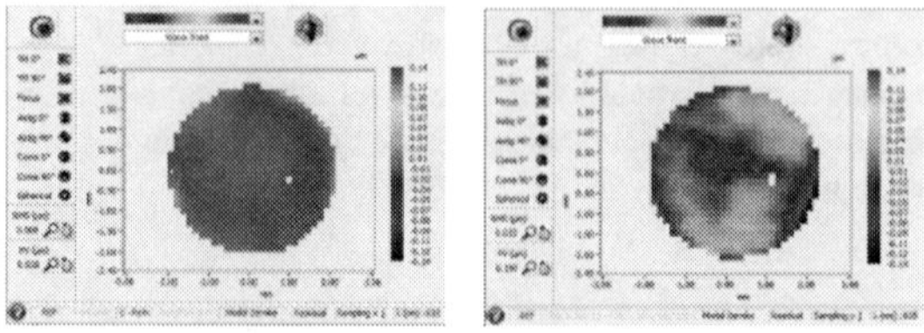

Fig. 4. (Left) Residual aberration in transmission at 635 nm WFE = 6 nm RMS and (right) in reflection at 1064 nm WFE = 32 nm RMS

2.2 Glass and flat optics

The SH technology is not based on interferometry. Thus, when measuring flat optics or glass, even for very thin samples, there is no parasite interferences due to Fabry Perot effect that would disturb the measurement. In the measurement bellow, it is possible to notice the effect of a black printing on the baking of the glass.

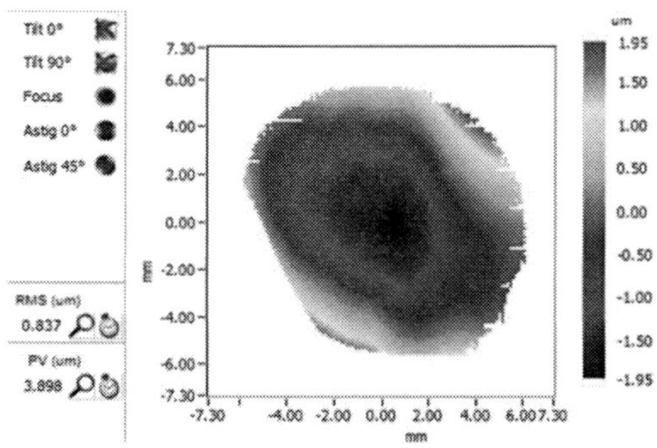

Fig. 5. SH wavefront map of a 2mm thick glass with a block print edge. WFE = 837 nm RMS

2.2 Optical assemblies

In some specific situations, characterization of an optical assembly requires a highly accurate measurement and a very compact sensor. In the case of large telescopes, the aberration is strongly dependant on its orientation because of the mechanical stress. That is why it is important to be able to design the metrology bench taking into account the orientation of the optical system.

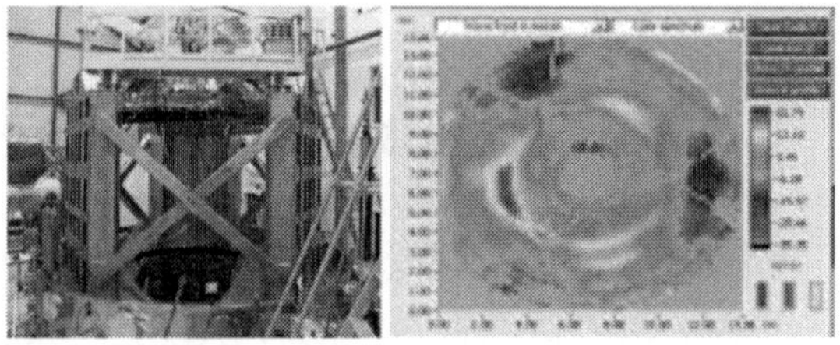

Fig. 6. Vertical optical characterization of Herschel telescope.

For the characterization of Herschel telescope, the HASO SH wavefront sensor was located below the telescope, and in this configuration, the wavefront map shows a non-neglectable trefoil aberration due to mechanical stress of the mirrors.

2.2 Aspheric lenses and freeform optics

With the advance of manufacturing techniques, freeform optics and aspheric lenses become more common. Thanks to its high dynamics – up to 3400 λ PV for HASO4 120 – and its absolute measurement capability, the SH wavefront sensor is an ideal tool for characterization of free from and aspherical optics.

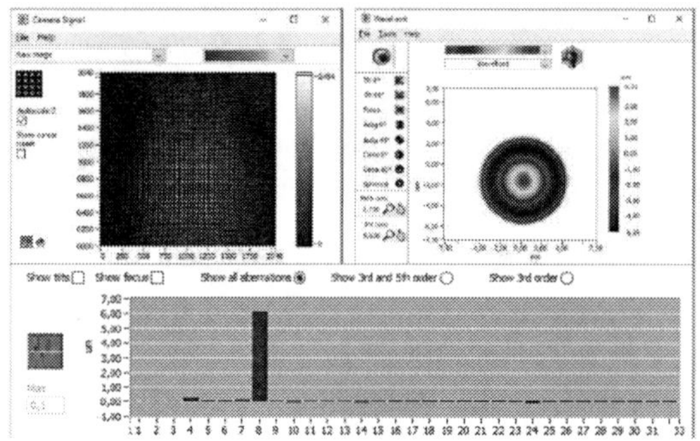

Fig. 7. Camera signal, wavefront and modal decomposition of a miniature aspheric lens.

2 Conclusion

Thanks to its features, the Shack Hartmann wavefront sensor is a good alternative for optics characterization in specific situations when the Fizeau interferometer fails at providing satisfying results.

Table 1 – Comparison between Shack Hartmann wavefront sensor and Fizeau interferometer.

SH sensor	Fizeau interferometer
Compact and portable	Table top on an isolated platform
Dynamic of tens of μm	Dynamic of a few μm
Limited spatial resolution	High spatial resolution
Multi-wavelength capability	Single wavelength
No reference beam	With a reference beam

References

1. Platt, Ben C.; Shack, R, *Journal of Refractive Surgery*. **17** (5): S573–7. PMID 11583233 (2001)

The Manufacturing of a Multi-surface Monolithic Telescope with Freeform Surfaces

Todd Blalock, Brian Myer, Brittany Cox, Jessica Nelson

Optimax Systems Inc., 6367 Dean Parkway, Ontario, New York 14519, USA

Abstract. Monolithic multi-surface telescopes combined with freeform optical surfaces provide improvements in optical performance in a smaller footprint as compared to systems with spherical surfaces, while providing superior mechanical stability to traditional telescope assemblies. Three different monolithic telescope concepts, in different configurations and optical performance were produced as proof of concepts. The results of the telescopes are presented as well as the unique manufacturing and testing challenges to produce these monolithic optical systems.

1 Advantages for monolithic optical systems

Monolithic systems offer the advantage of high stability as compared to traditional multi-element design. Alignment stability is crucial in high G-force applications such as airborne and spaceborne systems. The risk of element misalignment is greatly reduced by eliminating the need for opto-mechanical fixturing required to mount and position the optics.

2 Optical and mechanical design

In telescope systems off-axis freeform configurations are desirable when high contrast (no central obstruction) and reduced footprint is desired.[1,2] Recent improvements in optical freeform manufacturing have allowed for increased optical performance in off-axis optical systems which is particularly applicable in these monolithic structures.[3]

The designs of the three monolithic telescopes were a collaboration between Optimax Systems and NASA's Goddard Space Flight Center. Figure 1 shows the embodiment of all three monoliths. Figure 1 (a) is a two-freeform mirror solid glass prism, Figure 1 (b) is a three-freeform, diffraction limited, solid glass monolith, and Figure 1 (c) is a lightweight, two freeform mirror system. Both solid glass telescopes consist of a single blank of high purity fused while the light-weighted version uses a stable mirror grade glass substrate. All mirrored surfaces were given freeform prescriptions governed by XY polynomials. All three telescopes were designed with a field of view of $\pm$ 4.3° in Y-axis and $\pm$ 1.4° in X-axis at f/3.4. The two-freeform solid glass telescope shown in Figure 1(a) has four faces, faces 1 and 4 are respective plano entrance and exit faces and surfaces 2 and 3 are the freeform surfaces that provide the optical power. The entire telescope assembly is compact enough to fit within a single unit (1U) of a CubeSat payload with a weight of 1.3 kg. This design was not optimized to provide diffraction limited performance.

The next design iteration was the three freeform high-resolution solid glass telescope shown in Figure 1(b) which added a third freeform surface to provide diffraction limited performance. The additional glass material required to produce the increased optical performance contributes to its significant mass of ~2.3 kg.

The final design form in this study is the lightweight version shown in Figure 1(c). This light-weighted telescope contains the same optical prescription as the initial off-axis monolith design, however the light rays no longer propagate through any optical glass. This design reduced the weight by 60% to 0.42 kg.

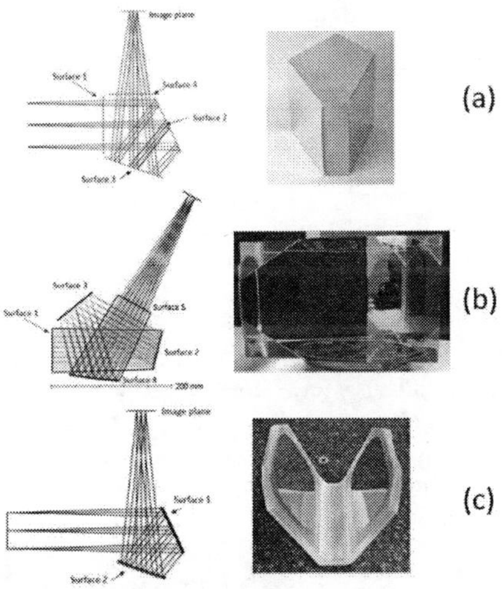

Figure 1. The three embodiments of monolithic telescopes shown with optical ray trace (left) and photograph of finished part (right). (a) Two freeform surface, solid prism, (b) diffraction limited, solid prism, and (c) two freeform, light-weight system.

* Corresponding author: tblalock@optimaxsi.com

3 Manufacturing and testing

3.1 Manufacturing considerations

All three monolithic telescope designs were polished on Optimax's custom and proprietary robotic freeform polishing platform [4, 5]. During the manufacturing process it is imperative that a consistent global coordinate system be defined to reference the freeform surfaces onto the monolithic structure. This coordinate system is usually based on easily accessible features built into the monolith. These features, or fiducials, are referenced during the manufacturing and metrology steps to ensure proper form of the freeform surface as well as its global position in 3D space. Figure 2 shows an example coordinate system for the lightweight monolith.

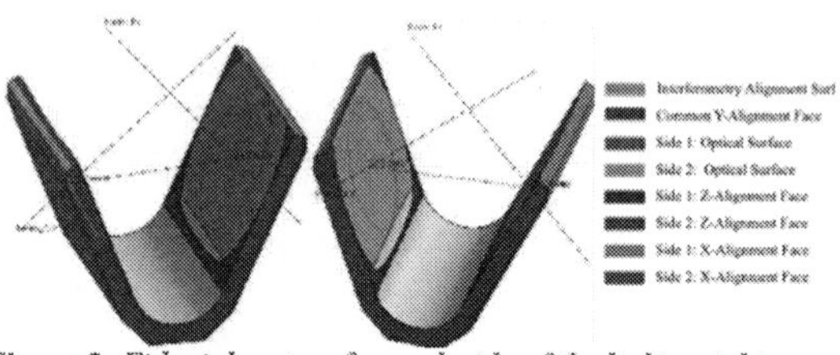

***Figure 2**: Fiducial system for each side of the light weight monolith.*

In the case of the lightweight monolith, additional manufacturing challenges included the design of miniaturized tooling. Due to the structure and shape of the lightweight monolith, there are issues with accessibility of the polishing tool to reach tight regions near the dihedral of the two surfaces. Due to the limited work space special care was taken to verify robot tool path to avoid collision between tooling and the part.

3.2 Testing considerations

Freeform metrology primarily relies on a two-phase approach using both contact metrology methods to locate the freeform surface within a reference system as well as a higher resolution measurement of the optical freeform surface. We mention several freeform metrology techniques used during the manufacturing and testing of the monolithic freeform telescopes that include two contact methods: coordinate measurement machine and a high resolution profilometer (Figure 3), as well as a non-contact method: deflectometry (Figure 4). [6]

***Figure 3**. CMM and UA3P at Optimax.*

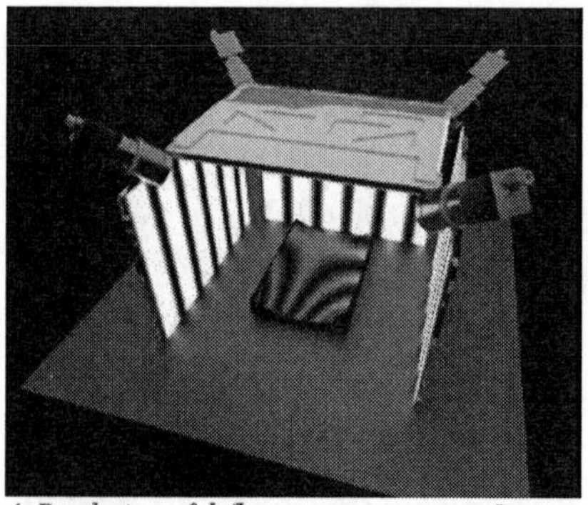

***Figure 4**. Rendering of deflectometry system at Optimax.*

Form and surface position is adjusted using CMM feedback down to a few micrometers of residual error. Then the surface can be evaluated, best fit, using deflectometry and UA3P. These can reveal MSF errors due to sub-aperture raster polishing, which must be removed via smoothing.

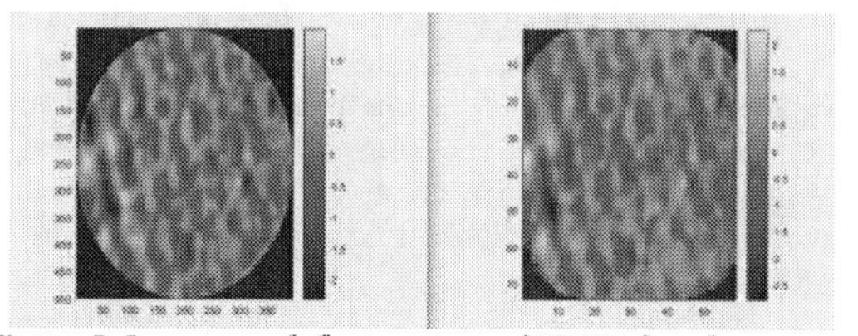

***Figure 5**. In-process deflectometry and UA3P data for lightweight monolith freeform surface.*

Conclusions :

Innovation in optical design, datum control, manufacturing and metrology are required to create high quality freeform monolithic telescopes for lightweight, flight applications.

References:

1. K.Fuerschbach et al., "Assembly of a freeform off-axis optical system employing three p-polynomial Zernike mirrors," Optics Letters **39**, 2896-2899 (2014).
2. J. M. Howard et al., "Improving the performance of three-mirror imaging systems with Freeform Optics," in Freeform Optics 2013, OSA Technical Digest, (The Optical Society, 2013), paper FT2B.6
3. J. L. Lawson et al., "Freeform monolithic multi-surface telescope manufacturing" in Optical Design Fabrication 2017 (The Optical Society, 2017), paper JTH1C.2
4. T. Lynch et al., "Deterministic form correction of extreme freeform optical surfaces" Proc. SPIE 9633, Optifab 2015, 96331F
5. D. Brooks et al., "Manufacturing of a large, extreme freeform, conformal window with robotic polishing," Proc. SPIE 10742, Optical Manufacturing and Testing XIII, 107420L
6. T. Blalock et al., "Metrology of Freeform Optics," Optical Design and Fabrication 2017 (The Optical Society, 2017), paper JTh4B.1

Final Figuring of Optical Components Surfaces Considering Deformations Introduced by Support System

Magomed Abdulkadyrov*, *Aleksandr* Semenov, *Nikolay* Dobrikov, *Aleksandr* Ignatov, *Vladimir* Patrikeev

JSC Lytkarino Optical Glass Factory (LZOS), Lytkarino, 140080, Parkovaya 1, Moscow region, Russia

Abstract. Membrane-pneumatic support systems for mirror mounting are often used at JSC LZOS during working surface tests while producing of large-dimensioned optical components of astronomical and space application. However, in some cases, it is necessary to support the mirror by glued interface elements. In such case, the difference in the surface shape between the standard and the technological support, which is considered during figuring, is calculated by means of modelling.

1 Introduction

In the process of figuring, the mirror is installed either on standard support, which is the best option, or, in some cases, if standard support is not available, on membrane-pneumatic support cell, which either supports the mirror uniformly, or imitates standard support cell. It allows both mirror figuring on grinding-and-polishing and finishing machine, and optical surface figure testing on vertical test bench, of which it is integral part. However, in certain cases, support system engineers design non-uniform efforts on interface elements of the mirror rear surface. In this case, the difference in the surface shape between the standard and the technological support, which is considered during figuring, is calculated by means of modelling.

2 Specific Features of Mirror Manufacturing Support

As an example, let us describe a number of projects on the production of large-dimensioned optical mirrors using membrane-pneumatic support.

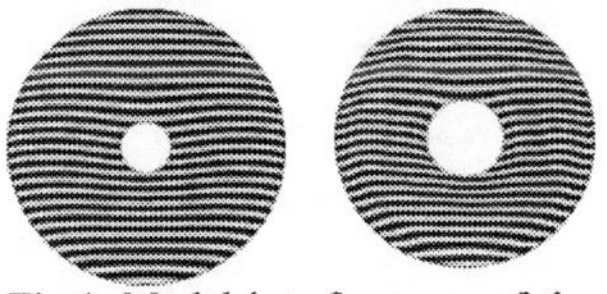

Fig.1. Model interferogram of the required wavefront (left) and the one actually obtained in the process of machining (right).

Under Contract with Belgium company AMOS, JSC LZOS fulfilled work on the production of the set of optics for Devasthal Optical Telescope (DOT) ARIES (Aryabhatta Research Institute of Observational Sciences, India) [1]. The telescope was produced by AMOS. The primary mirror is a meniscus. The mirror diameter is 3700 mm, vertex radius is 14639 mm (F/1.96), asphericity is 111 μm.

The estimated manufacturing support was designed in such a way that the efforts on the points of mirror support are different. Both LZOS and AMOS specialists performed modelling of the supported state. The calculation results matched each other. Modelling results showed that locating of membranes in the points of standard support cell efforts induces errors on the surface, mainly, defocus and 3rd order spherical aberration. The rest of the errors, «traces» from support elements, are small and negligible, since they are included in the mirror deformation calculations both in the standard support cell and in the manufacturing support cell. Decision was taken to subtract in the process of mirror machining the error induced by the manufacturing support cell, together with AMOS specialists and thus obtain the surface figure compliant with the standard support cell. Defocus does not influence the mirror parameters, since it introduces error within radius of only 0.07 mm, and spherical aberration should be considered during mirror final polishing. In order to do this, it is necessary to obtain wavefront map with the required spherical aberration, which will be removed upon installation of the mirror on the standard support cell. Focus and 3 terms of Spherical aberration as given by the equation hereunder:

$$-207\cdot(6\rho^4-6\rho^2+1)+42.8\cdot(20\rho^6-30\rho^4+12\rho^2-1)-17.6\cdot(70\rho^8-140\rho^6+90\rho^4-20\rho^2+1)$$

Interferogram of the required wavefront is showed in Fig.1, on the left. The parameters of such surface are:

RMS (W) = 0.133λ; PV (W) = 0.458;

RMS SFE = 42.0 nm

By subtracting of this wavefront from the obtained one in the process of the mirror surface testing we remove the difference between support in the standard support cell and in the manufacturing membrane-pneumatic support cell and perform machining to meet the specification requirements. Possible errors in the mirror supports difference modelling in the process of

* Corresponding author: lastro-am@lzos.ru

machining and operation are compensated by the active system of impact on the primary mirror.

Another example – under Contract with AMOS, JSC LZOS is producing optics for DAG Telescope (Doğu Anadolu Gözlemevi) [2]. The primary concave hyperbolic mirror has diameter 4 m, vertex radius Rc = 14420 mm ± 12 mm (F/1.80), conic constant K= - 1.006574± 0.0004, asphericity is 153 μm on the full diameter and 139 μm in the light zone.

Modelling results showed that locating of membranes in the points of standard support cell efforts induces errors on the surface, mainly, defocus, astigmatism, triangular coma and spherical aberration. The rest of the errors, «traces» from support elements, are small and will be removed in the process of figuring. Defocus is 1.5 μm and does not influence the mirror parameters, since it introduces error within radius of only 0.12 mm, however, such error can also be considered during figuring.

By subtracting of this wavefront from the obtained one in the process of the mirror surface testing we remove the difference between support in the standard support cell and in the manufacturing membrane-pneumatic support cell and perform machining to meet the specification requirements.

The view of wavefront and the corresponding interferogram, which should be obtained on membrane-pneumatic cell, are showed in Fig. 2-3, and the errors of wavefront are showed in Fig. 4.

The accuracy of positioning of the mirror support system interface elements is also important. In particular, positioning of the axial interface elements of DAG M1 mirror was performed with a higher accuracy, 0.02-0.03 mm instead of 1.2 mm. The purpose of this was to decrease considerably the errors induced by the manufacturing support cell influence on the mirror deformations in the process of its installation on the memebrane-pneumatic support. Calculations of the mirror deformations in case of its shift with respect to the cell by 0.5 mm, 1 mm, as well as in case of error in the axial interface element positioning by ±0.5 mm and ±1 mm were carried out. Influence of membrane shift with respect to the nominal position within 1-2 mm is negligible, since membrane mounting pad of diameter 80 mm has the same influence on the interface element as without the shift; at the same time, shift of the interface element causes the effort to be applied to the mirror in the shifted place, and this would have considerable influence on wavefront form. Within the telescope, this effect is compensated by active influence on the pad, while static membrane-pneumatic support lacks such compensation. Calculations showed that the axial interface elements should be glued with accuracy <0.5 mm from the nominal position, which is exactly the way it was done.

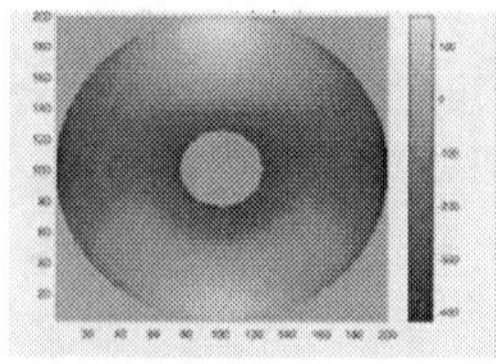

Fig. 2. Wavefront form in case of mirror support on manufacturing support cell.

Fig. 3. Model interferogram of the required wavefront and parameters of wavefront deviations.

	RMS	P-V
All	0,348	1,823
- Tilt	0,348	1,823
- Focus	0,348	1,823
- Astigmatism	0,168	0,770
- Coma3	0,348	1,823
- Coma5	0,348	1,823
- Triangular	0,345	1,749
- Quadratic	0,348	1,820
- Spherical	0,315	1,702
Local	0,068	0,344

	Coefficient
Focus	0,001

	Coefficient	Angle
Tilt	0,000	-0,060
Astigmatism	0,749	0,016
Coma3	0,002	-5,992
Coma5	0,003	169,632
Triangular	0,126	0,023
Quadratic	0,014	-44,987

	Coefficient
Spherical (0)	-0,790
Spherical (2)	6,328
Spherical (4)	-15,219
Spherical (6)	15,784
Spherical (8)	-6,475

Fig.4. Parameters of the required wavefront.

In the course of figuring of such surface, it is possible either to subtract the required wavefront from the one obtained during testing, or to produce CGH, which considers the form of the required wavefront showed in Fig. 2-3. In this case, we would actually test aspherical mirror of arbitrary surface figure as spherical one. Only in such case, it is necessary to carry out fine alignment of the hologram with respect to the component by matching its angular position with respect to the tested component with the mirror vertex position.

References

1. A.P. Semenov, Proc. SPIE, 8450, 84504R-1 (2012)
2. M. A. Abdulkadyrov, A. P. Semenov, S. P. Belousov, N. M. Vladimirov, A. N. Ignatov, V. E. Patrikeev, V. V. Pridnya, A. V. Polyanchikov, R. K. Nasyrov, O. Pirnay, C. Flebus, C. Yesilyaprakd, O. Keskine, Proc. SPIE 10706, 1070631-1 (2018)

 https://doi.org/10.1051/epjconf/201921507001

Analysis of residual errors during computer controlled polishing

Simon Killinger[1,*], *Alexander* Haberl[1], and *Rolf* Rascher[1]

[1]Deggendorf Institute of Technology, Dieter-Görlitz-Platz 1, 94469 Deggendorf

Abstract. In computer controlled subapertur polishing the formation of mid spatial frequency errors (MSFE) needs special attention. In this work the formation of MSFE in feed direction is investigated using the ADAPT tool from Satisloh.

1 Introduction

In previous projects computer controlled polishing (CCP) was found to be a suitable method for correcting form errors in precision optics. However, CCP methods tend to produce mid spatial frequency errors (MSFE) on the surface, which is especially critical when the polished optics are used for high power laser applications [1, 2]. Using the ADAPT tool from Satisloh the formation of MSFE in feed direction is in depth investigated by polishing a complete surface. This also extends the work from [3] where the shape of the removal function is analyzed using one path.

2 Experiment setup

The ADAPT tool consists of two parts: A polyurethane (PU) polishing foil and a PU pin. The pin is brought to a specific radius and the polishing foil is bent over this pin. The tool is then dressed in a MCP 250 machine from OptoTech.

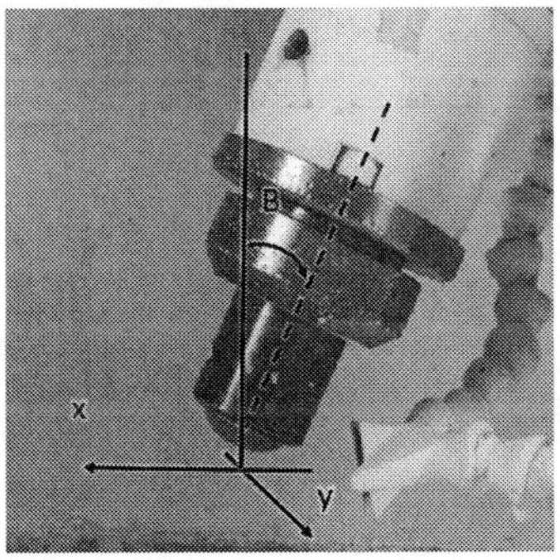

Figure 1. The Satisloh ADAPT tool as mounted in the OptoTech MCP 250 [4]

As seen in Figure 1 the tool rotates around its axis of symmetry and is tilted by the angle of precession B. The tilted tool is then moved across the workpiece in a meander pattern. The meander pattern consists of a reciprocating movement in x direction and a path distance between each path in y direction.
The resulting surface is afterwards interferometrically measured using a VeriFire XP/D by Zygo and the Zygo MetroPro 8.3.4 software for evaluation. To get the resulting spatial frequencies, four randomly positioned profiles are placed on the measured surface in feed direction using MetroPro. From these profiles the amplitude of the main frequency is noted and the mean is calculated.

3 Experiment results

The experiment shows, there is a correlation between the resulting spatial frequency error f in feed direction and the ratio between the feed speed v_F and the rotational frequency n as shown in Equation 1.

$$f = \frac{n}{v_F} \tag{1}$$

Concluding from these results, the irregularities of the polishing tool depict on the surface after each rotation. By varying the parameters path distance, compression and precession angle the amplitude of the resulting main frequency can be lowered or raised. Whereas by adjusting the ratio from Equation 1 above a certain threshold the amplitude of the resulting frequency falls to an almost non measurable value. With the used parameters this threshold ranges at about five tool rotations per 1 mm feed length. This coherence is surveyed in [4] and this coherence can also be seen in grinding as described in [5].

4 Simulation

In order to recreate the observed pattern a simulation is carried out using Python. The simulation is based on the assumption that the MSFE results from irregularities on the surface of the polishing foil which imprints on the substrate surface when spiraling across the surface with the feed and rotational frequency of the tool. Hence these

*e-mail: simon.killinger@th-deg.de

irregularities are approximated as a virtual TIF spiraling across the surface as seen in Figure 2.

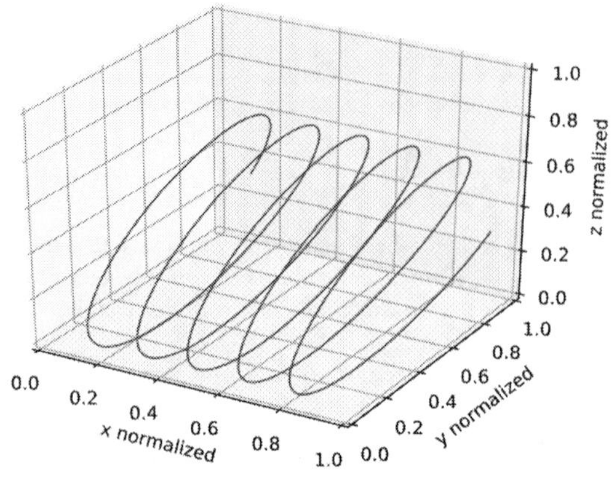

Figure 2. Normalized depiction of the spiraling movement of the irregularities on the ADAPT tool surface cf.[4]

The resulting tool path for the virtual TIF across the surface is shown in Figure 3. The tool path is generated by shifting the spiral from Figure 2 by the compression of the tool in z direction. The positive part of the spiral is then cut off. The remaining part of the spiral defines the tool path on which the virtual TIF removes material from the substrate in one path. In the respective path distance in y direction further paths are lain. These paths are also shifted by a small arbitrary offset dx in x direction. This arbitrary offset ensures that the spiral sections do not completely align. Thus the reality is emulated where the fast rotation over the slower feed speed randomizes the position of the respective irregularities on the tool surface.

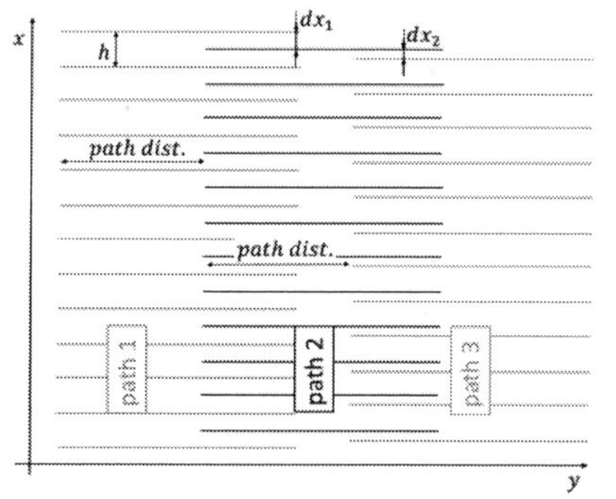

Figure 3. Normalized depiction of the spiraling movement of the irregularities on the ADAPT tool surface

The removal in CCP can be calculated as the convolution of a TIF with a dwell time and since the dwell time remains constant the tool path can be used. Hence the virtual TIF is convoluted with the beforehand generated tool path. The result is shown in Figure 4. This so simulated removal can be imported in MetroPro to measure the amplitude of the resulting main frequency in the same way as it was done for the experiments.

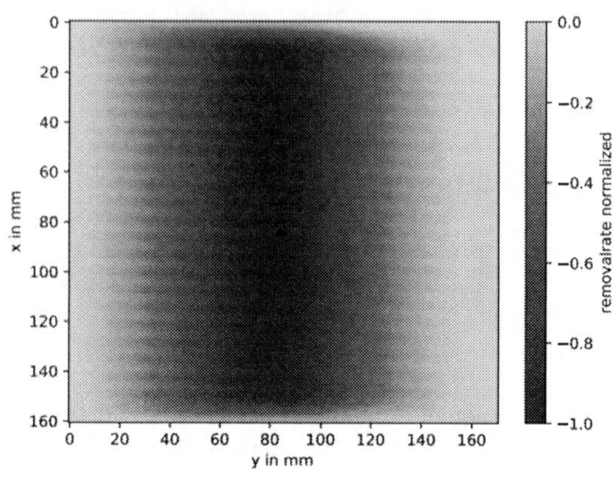

Figure 4. Simulated removal generated by the virtual TIF on the calculated tool path

5 Simulation results

By varying the virtual TIF the results from the experiment can be approximated hence the effect from Equation 1 may be predictable. Also the threshold as seen in the experiment can be replicated. This threshold may result from the formation of a plateau when the rotations per mm reach a certain value. The arbitrary offset in x direction also has an impact on the amplitude of the resulting spatial frequency.

References

[1] Bliedtner, Jens and Graefe, Günter *Optiktechnologie* (Carl Hanser Verlag, München, 2010)

[2] Zhong, Bo; Huang, Hongzhong; Chen, Xianhua; Deng, Wenhui and Wang, Jian **Modelling and simulation of mid-spatial-frequency error generation in CCOS**, (J. Eur. Opt. Soc.-Rapid Publ. (Journal of the European Optical Society-Rapid Publications), 2018)

[3] Schneider, Robert; Haberl, Alexander and Rascher, Rolf, **Polishing tool and the resulting TIF for three variable machine parameters as input for the removal simulation**, (Proc. of 4th European Seminar on Precision Optics Manufacturing, 2017)

[4] Killinger, Simon; Liebl, Johannes and Rascher, Rolf **Mid-Spatial-Frequency-Errors in feed direction occurring in ADAPT Polishing**, (Proc. of 6th European Seminar on Precision Optics Manufacturing, 2019)

[5] Pohl, Mario; Börret, Rainer; Rascher, Rolf and Kukso, Olga, **Simulation of MSF errors using Fourier transform**, (Proc. of 5th European Seminar on Precision Optics Manufacturing, 2018)

 https://doi.org/10.1051/epjconf/201921507002

Fields of application for laser beam polishing of optical glasses

Anne-Marie Schwager*, *Jens* Bliedtner, *Kerstin* Götze, *Michael* Seiler, *Tino* Ritz, *Christoph* Letsch

Ernst-Abbe-University of Applied Sciences Jena, Carl-Zeiss-Promenade 2, 07745 Jena

Conventional manufacturing of freeform optics is expensive and elaborate. The laser beam as a polishing tool offers the possibility to produce optical surfaces fast and effective. [1] [2]
The aim of new research on laser beam polishing of optical glasses is to achieve a polished surface finish irrespective of the surface geometrics. The carbon dioxide laser is applied to process pre-machined specimen with different roughness's. During the polishing process an infrared camera is necessary to monitor the temperature on the surface. An annealing process reduces the process-related tensions. In addition, the cleaning process is investigated to minimize the surface damages. To characterize the surface topographic a white-light interferometer and a stylus instrument are applied.
Laser beam polishing of minerals glasses such as N-BK7®, BF33® and fused silica enables surface roughness of less than Ra < 5 nm by form maintenance. By means of self-design setup, the adaption of laser beam polishing on glass cylinders, lenses or freeform is possible. As a result a parallel laser beam with a diameter of 12.65 mm is applied to process tilted surfaces up to 45°. With a laser parameter modulation, the angle-dependent absorption depending on the intensity of the laser radiation is considered.
The picture shows a laser beam polished lens surface made of optical glass N-BK7®.

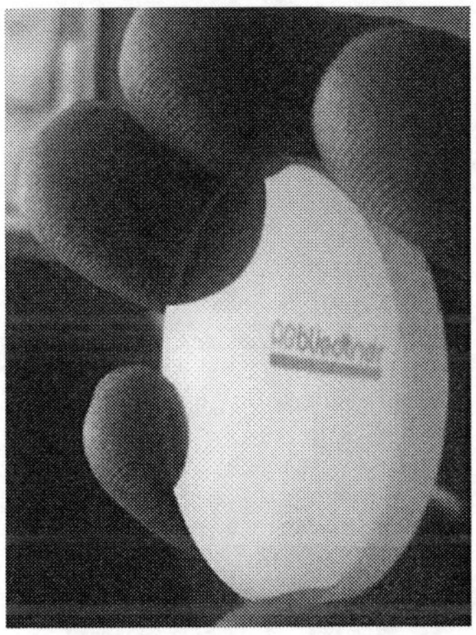

1. Bliedtner, J., Müller H. and Barz A.: "Lasermaterialbearbeitung, Grundlagen- Verfahren- Anwendungen- Beispiele " Carl Hanser Verlag , 2013. - ISBN 978-3-446-42168-4.
2. Hecht, K.: "Entwicklung eines Laserstrahlpolierverfahrens für Quarzglasoberflächen", Dissertation, Ilmenau, Universitätsverlag Ilmenau

This work has been partially funded by the German Federal Ministry of Education and Research as part of the funding program Forschung an Fachhochschulen. Funding reference: 13FH003IA6 "OpTec4.0". The authors wish to thank the project partners for supporting the investigations.

* Corresponding author: anne-marie.schwager@eah-jena.de

EPJ Web of Conferences **215**, 08001 (2019)
EOS Optical Technologies https://doi.org/10.1051/epjconf/201921508001

Approaches For A Destructive Measurement Method Of Subsurface Damages

Michael Seiler[1,*], *Lukas* Tianis[1], *Jens* Bliedtner[1], *Mario* Berlinger[2] *Steffen* Gürtler[2]

[1]Ernst-Abbe-Hochschule Jens, Department SciTec, Carl-Zeiss-Promenade 2, 07745 Jena, Germany
[2]Bühler Alzenau GmbH, Föpplstraße 9, 04347 Leipzig, Germany

Abstract. In optical manufacturing, the depth of subsurface damages (SSD) depicts an important quality aspect of optical elements. Nowadays, the investigation methods of the damage zone are mostly destructive and allow a sampling of the substrates only. This leads to a lack of statistical reliability. In this paper, one aim is to develop an easy and cheap procedure to evaluate the probes. Therefore, a spherical polishing and chemical etching is introduced. By means of imaging processing software the acquired data is analysed, which eliminates the operator influence und increases the reproducibility. Applying binarizing and shape detection algorithms the SSD structures is detected and the number of pixels per structure is calculated. The percentage of SSD over the depth fit with an exponential curve displays a maximum depth as a single value. The development of a method to choose the best threshold value represents an approach to quantify the uncertainty of the resulting SSD-depth.

1 Introduction

The conventional optical manufacturing processes, such as grinding and lapping, leave a damaged layer, which includes micro cracks. These subsurface damages (SSD) are one of the important quality characteristics of optical elements. One major interest is the evaluation of the depth of the damaged zone which determines the material removal for the following polishing pro-cess and, thereby, the time and cost. Nowadays, most of the used techniques to investigate SSD are destructive and only allow sampling to test the processed substrates [1]. This leads to a lack of statistical reliability of the measurement process.

2 Experimental results

The development of the new procedure shown in **Fig. 1** for the preparation of samples of N-BK7, SF6 and HPFS7980 by spherical polishing and chemical etching was the first aim. The elimination of operator influences and increase of reproducibility of the evaluation process was another. To do so, a self-developed image processing software analyses the microscope photographs of the pre-pared surface taken on random positions of the etched surface. Binarizing and shape-detection algorithms are the key functions to detect SSD structures.
Fig. 2 displays some typical SSD-structures after etching. The SSD are marked in red. In order to get the percentage of SSD, the number of pixels is calculated.

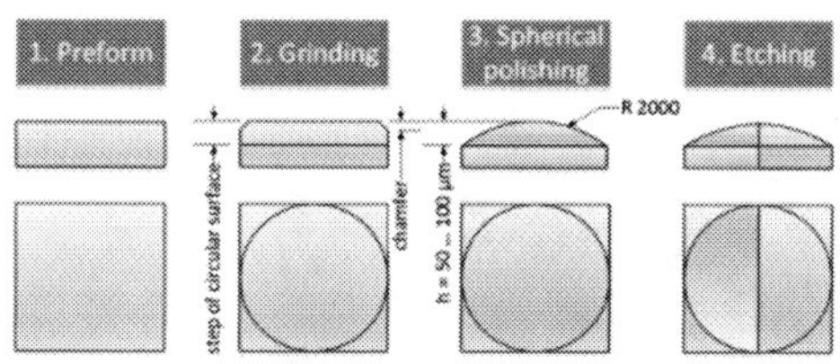

Fig. 1. Spherical Polishing process.

The spherical shape of the specimen is measured by a contact stylus profilometer. Due to the single profiles the depth of removal at any distance from the center of the specimen is provided.

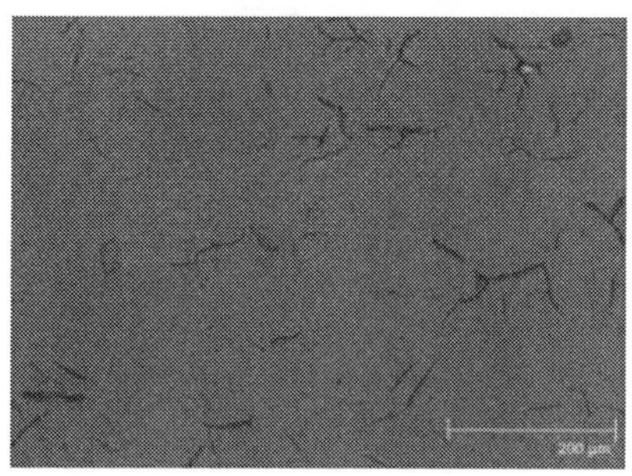

Fig. 2. SSD evaluation (marked red) with digital microscope.

* Corresponding author: michael.seiler@eah-jena.de

To investigate the SSD itself, laser-scanning microscopy (LSM) provides images of the topography. Images of the unetched surface identify peaks above the surface level.

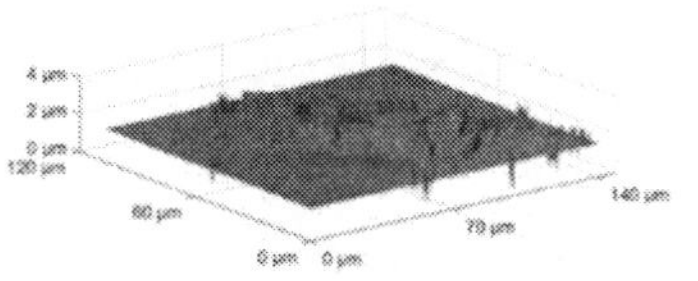

Fig. 3. LSM mapping of an unetched surface.

After wet etching, these structures increase in lateral dimension. At the contour map these structures display as valleys. It is presumed that the unetched structures create reflections of the laser, which lead to inappropriate measurement result. After etching, the structures are wided and the reflection are eliminated. Consequently, the examined structures are subsurface-damages. To further examine the SSD, a digital light microscope acquires the images, because there is no need to gain data of the height of the surface. Light microscopy can detect SSD with sufficient precision. An examination of the different possibilities to recognize SSD on etched and non-etched specimen by taking images of the transition zone between the two parts of the specimen is performed.

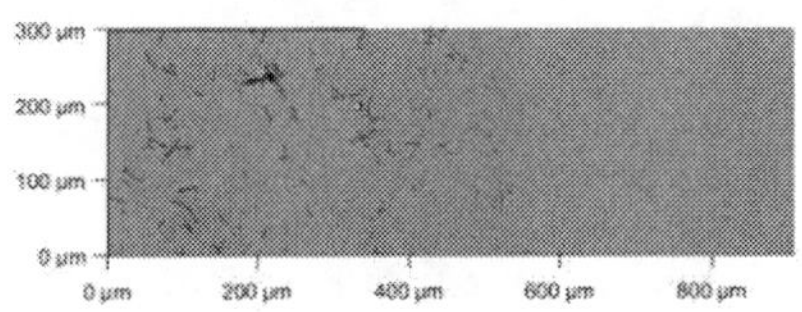

Fig. 4. Comparison between unetched and etched region on a N-BK7 surface.

Fig. 4 shows the transition zone. SSD on the etched part (left) occur as dark structures whereas on the unetched part (right) no SSD can be identified. Summarized, an etching of the specimen is indispensable the evaluate SSD.

By linking the surface profile with the positions of the microscope photographs and the calculated percentage of SSD, a plot of SSD-density over the depth of removal is created. As shown in Fig. 5, the data can be described by an exponential fit shown in Eq. (1). $\%_{SSD}$ is the percentage of SSD d_r, the depth of removal and a, b the coefficients of the fit.

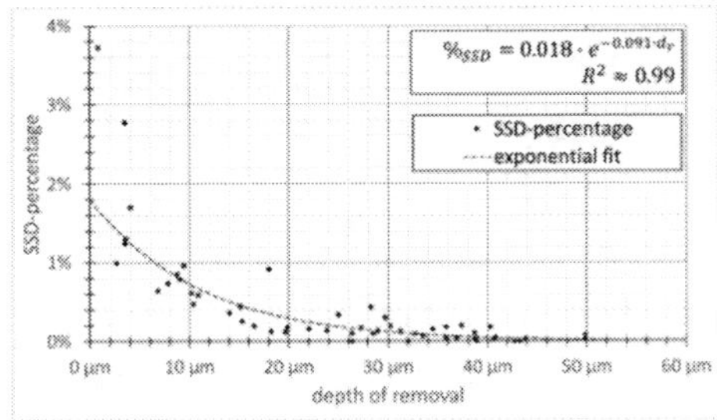

Fig. 5. Exponential Fitting of the SSD-percentage over depth of removal.

$$\%_{SSD} = a \cdot e^{b \cdot d_r} \tag{1}$$

After executing the test for outliers and eliminating these data, the coefficient of determination is in general higher than 0.95. Threshold values applied on the fitted curve describe the maximum depth of SSD as a single value. For that, Eq. (1) is resolved to x and the threshold value is used for f(x) in Eq. (2). the fit.

$$d_r = \ln \frac{\%_{SSD}}{a} \cdot \frac{1}{b} \tag{2}$$

The threshold value has an immediate influence on the statistical reliability and, therefore, needs to be chosen prudently.

3 Summary and Outlook

A Comparison of the different glasses and the examination of the whole measurement process will describe its suitability. The development of a method to choose the best threshold value represents an approach to quantify the uncertainty of the resulting SSD-depth. Additionally, an improvement of the software should be sought to detect SSD with high precision.
Furthermore, additional materials will be tested. The detection of the SSD shape can be enhanced with object detection method and further machine learning.

Acknowledgements

Work is performed under the auspices of the German Federal Ministry for Economic Affairs and Energy under funding reference ZF4039605PR6.
The authors would like to thank Dietmar Gräfe and Sebastian Henkel for processing the samples.

References

1. Y. Li, J. Wang, Q. Xu and Y. Guo, Journal of the European Optical Society - Rapid Publications, (2011)

Measurement and correction of two-sided freeform optical elements with combined tactile-optical metrology equipment

Johannes Hartung[1,*], *Henrik* von Lukowicz[1,2], *Mathias* Rohde[1], *Knut* Kleinbauer[1], *Nils* Heidler[1], and *Stefan* Risse[1]

[1]Fraunhofer Institute for Applied Optics and Precision Engineering IOF, Albert-Einstein-Strasse 7, 07745 Jena, Germany

[2]Institute for Applied Physics IAP, Friedrich-Schiller-University Jena, Max-Wien-Platz 1, 07743 Jena, Germany

Abstract. Freeform optical elements are state of the art for several years to fabricate very high performance optical systems with the necessity of, e.g., strong folding in mirror system or correction of typical asymmetric aberrations in mirror systems as well as lens systems. For freeform mirror systems, in particular for metal mirrors, the metrology is well understood and iterative process chains are well established. For transmission elements with a freeform surface on both sides, manufacturing, metrology, and correction for both sides in a parallel manner is quite difficult. The article presents a method to measure such an optical element and correct it with a well-defined correction step to have both sides in a well-defined position to each other.

1 Introduction

Freeform optics are state of the art for several years now. They are often used in systems where compactness or the correction of asymmetric aberrations is important. Since freeform surfaces are more difficult in manufacturing and measurement, they are more expensive and therefore usually only used for high performance optics. For freeform metal mirrors a fabrication, metrology, and correction process chain is well-established from MIR to VIS wavelengths. The advantage of metal mirrors is that it is possible to manufacture references directly on the mirror substrate near the optical surface and that in mirror systems chromatic aberrations play no role at all [1, 2]. For freeform lenses of high performance transmission optics, the situation is more complicated than for metal mirrors. First of all it is not possible to manufacture metrology references directly on the lens substrate. Further, lenses have usually not enough edge area to be usable for references. The most difficult part for manufacturing of freeform lenses is that if every side of the lens is a freeform, it is necessary to measure the two sides relative to each other to quantify the optical error introduced by the misalignment. In a second step, these errors have to be corrected while it is necessary to separate form and position error in a well-defined way. The article introduces a metrology equipment to make the measurement of form and position error possible by using different metrology methods and to later correct both sides of a freeform lens with respect to form and position.

2 Preliminaries

A lens is divided into two optical surfaces and an outer boundary. The two surfaces in the present case are freeforms from the viewpoint of manufacturing. This means there are more than two independent axes for machining necessary to manufacture such surfaces. The outer boundary consists of mounting structures and reference features for later measurement or mounting within a frame structure. For glassy elements it is difficult to manufacture reference spheres into the outer boundary. The only reference features available are angular flats, plane surfaces, and circular structures for centering (see, e.g., fig. 1).

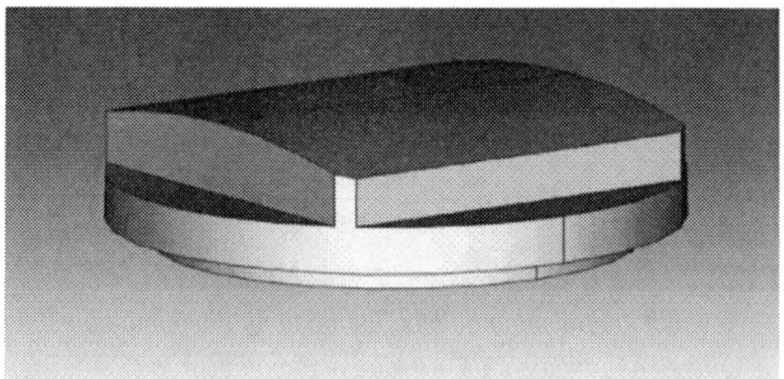

*Fig. 1 **Plano** surface for alignment and circular boundary for centering. The **plano** surfaces in the XY plane can be used for detilt.*

The overall challenge is to measure the reference structures at the boundary with respect to the optical surfaces to have some geometrical features as a first part in the measurement chain. From these measurements, it is possible to measure the positions of the optical surfaces

* Corresponding Johannes Hartung: johannes.hartung@iof.fraunhofer.de

and separate form and position error from each other. The available measurement machines, which provide the necessary high measurement accuracy, are a coordinate measurement machine (Zeiss Mikura), a 2.5D profilometer (Panasonic UA3P), and an interferometer (Zygo DynaFiz). Since there is no possibility to get a common reference system for all three measurement machines, it is necessary to develop a measurement equipment (see fig. 2).

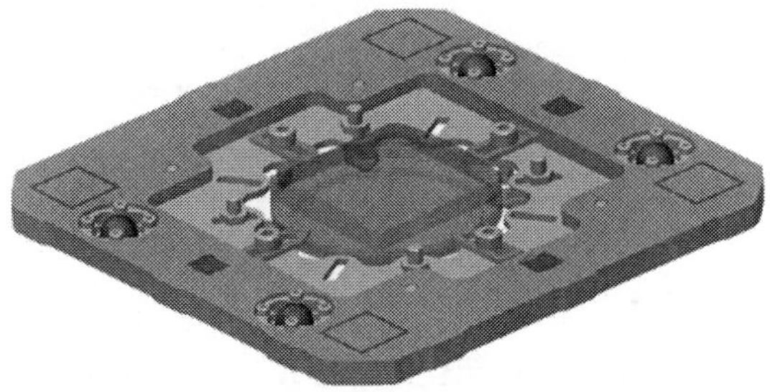

Fig. 2 Measurement equipment – CAD design. Red are the surfaces for alignment in different measurement machines. Dark yellow are the Rubin spheres. In the middle there is the inlay with a demonstrator optics for the project.

This equipment [3] is designed as a kind of frame which has several tactile and optical references on its exterior structure and a CNC machined inlay for fixing the lens itself. The interior inlay can be easily rebuilt for another lens whereas the outer structure needs to be designed for a certain size and accuracy of the measurements. The different reference structures are:

- 4 plano surfaces on each side for removing tip/tilt in front of an interferometer
- 4 cylindrical CGH references for fixing the position of the CGH in relation to the measurement equipment on each side
- 4 Rubin spheres built into the frame which are useful for finding a common coordinate systems for both sides

The final goal is to arrange all measurement data into one common coordinate system which leads to a high data analysis effort. But afterwards it is possible to perform a post-polish process for glassy materials with the help of Magnetorheological finishing (MRF). The relative position and the form of each of the freeform surfaces can therefore be generated with a high accuracy depending on the quality of the reference features.

For the MRF polishing part, it is necessary to manufacture another manufacturing equipment to fix the freeform surfaces with all six degrees of freedom such that the polishing indeed corrects for the positional as well as the form errors of the optical freeform element. This is necessary since the measurement equipment cannot be adapted for the integration into the MRF machine.

3 Results

Within the project the first data was generated by measuring a lens which has one plano side and another convex cylinder side with curvature radius of 40 mm along the X axis.

For this test part it was verified that it is possible to arrange profilometer as well as coordinate measurement data into one coordinate system from which it is possible to achieve a form and position correction onto that substrate (see fig. 3).

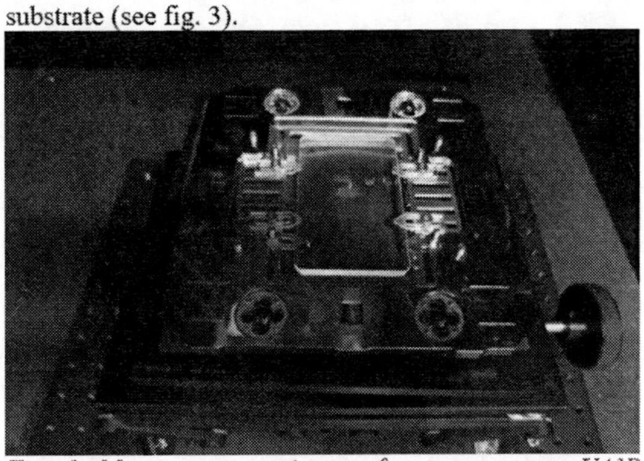

Fig. 3 Measurement equipment for test part on UA3P profilometer.

Fig. 4 Demonstrator element on MRF for form correction.

Additionally, an interface for the MRF correction of the demonstrator was manufactured (see fig. 4) to verify the positioning step using reference elements.

The authors would like to thank Matthias Beier for enlighting discussions, Wilko Fuhlrott for diamond turning of the measurement equipment, and Stephanie Hesse-Ertelt for language corrections. The research is funded in project Free4Optics under grant number 2015 FE 9148 by the Thüringer Aufbaubank (TAB) within the European Regional Development Fund (ERDF).

References

1. S. Scheiding, C. Damm, W. Holota et al., Proc. SPIE **7739**, 773908 (2010)
2. M. Beier, J. Hartung, T. Peschel et al., Appl. Optics **54**, 3530 (2015)
3. Patent 2018 F61223, Halterung zur Fixierung und Referenzierung von mit einer Messvorrichtung an mehreren, nicht in einer gemeinsamen Ebene angeordneten Oberflächen in einer Einspannung zu vermessenden Bauteilen

https://doi.org/10.1051/epjconf/201921508003

Investigation of Non-Uniformity of Classically-Polished Fused Silica Surfaces via Laser-Induced Breakdown Spectroscopy

Christoph Gerhard[1,*] and *Jörg* Hermann[2]

[1]University of Applied Sciences and Arts, Faculty of Natural Sciences and Technology, 37085 Göttingen, Germany
[2]Aix-Marseille University, CNRS, LP3, 13009 Marseille, France

Abstract. In this contribution, the surface uniformity of classically-manufactured fused silica windows was investigated via laser-induced breakdown spectroscopy. It is shown that for all investigated samples a comparatively high aluminium content was found at the edge of the surface with respect to its centre. This contamination can be attributed to residues from lapping and polishing agents and leads to a mentionable non-uniformity of the surface in terms of chemical composition and optical properties, respectively.

1 Introduction

The quality of an optics surface regarding its contour accuracy or cleanliness is extensively covered by the pertinent standards such as ISO 10110. This also applies to the homogeneity of the bulk material. However, the surface uniformity, i.e. the spatial variation of its chemical and optical properties, is not specified. This point becomes of interest for the application of high performance coatings where uniformities in final surface reflection or transmission over the surface of 0.1 to 0.5% are required. In this context, the homogeneity of the coating material and the coating thickness are strictly specified and the stability of the coating process is meticulously controlled, but less attention is usually paid to the uniformity of the substrate surface. Against this background, the homogeneity of fused silica window surfaces was investigated in the present work where the focus was on the polishing agent aluminium oxide.

2 Materials and methods

The investigated samples were classically-manufactured commercial samples made of Suprasil 2B with brightly polished surfaces (P3-polish). The chemical analysis of the surfaces was carried out via laser-induced breakdown spectroscopy (LIBS), a powerful tool for the analysis of glasses [1]. As described in detail in [2], measurements were performed at various locations on the sample surface in order to obtain a cross section of the chemical composition. In addition to such cross-sectional detection, the edge and the centre of the samples were investigated by in-depth LIBS where single laser shots were successively applied. Since no other polishing agent materials such as cerium or zirconium were detected, the distribution of aluminium on the sample surfaces was measured. Here, the LIBS analysis was performed by observing the Al I 396.15 nm resonance line.

3 Results and discussion

As shown in Figure 1, the lateral distribution of aluminium over the sample surfaces features the same qualitative behaviour for all investigated samples: A comparatively high aluminium contamination is found at the border areas or edges of the samples whereas low or even no contamination occurs at the centres.

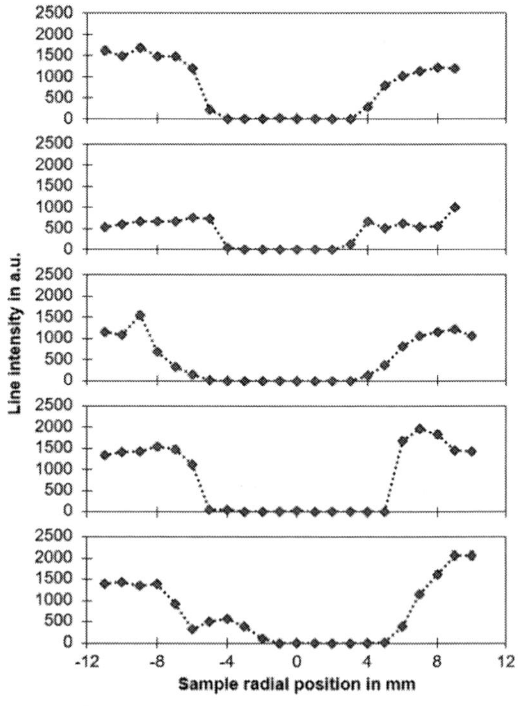

Fig. 1. Intensity of the Al I 396.15 nm transition vs. sample radial position of different commercial fused silica plates.

* Corresponding author: christoph.gerhard@hawk.de

For the edges, a mean aluminium mass fraction of about 1,000 ppm (0.1%) can be estimated based on the measurements [2]. This high aluminium content cannot be attributed to the glass bulk material since the investigated type of fused silica contains a negligible aluminium content of <10 ppb according to the glass manufacturer's data sheet. It can rather be explained by the contamination of the surface by aluminiferous lapping and polishing agents such as corundum in the course of the manufacturing process.

The observed lateral distribution is another point of great interest. It could be explained by the fact that in classical optics manufacturing, the surface is usually shaped slightly concave with respect to the final radius of curvature by each single previous process step. As a result, the subsequently applied tool rests on the border of the component [3]. Such edge support leads to an extremely stable self-guidance of any following grinding, lapping or polishing step. However, it also increases the strength and time of interaction of the tool and the border area of the work piece. In comparison to the centre, the tool's impact is thus much higher at the edge, leading to a larger density of micro cracks where contaminants from working materials such as – in the present case – aluminium oxide can accumulate.

This assumption is also supported by the performed in-depth measurements where a high contamination and penetration depth of aluminium at the edge with respect to centre was verified as shown in Figure 2.

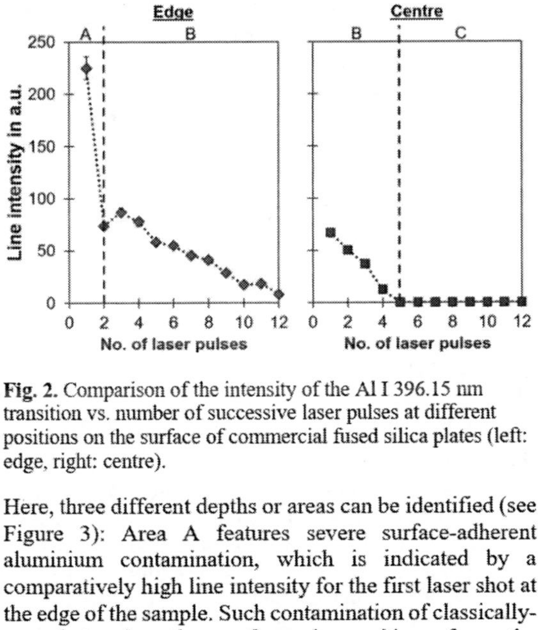

Fig. 2. Comparison of the intensity of the Al I 396.15 nm transition vs. number of successive laser pulses at different positions on the surface of commercial fused silica plates (left: edge, right: centre).

Here, three different depths or areas can be identified (see Figure 3): Area A features severe surface-adherent aluminium contamination, which is indicated by a comparatively high line intensity for the first laser shot at the edge of the sample. Such contamination of classically-manufactured optics surfaces by residues from the polishing agent aluminium oxide was also reported in [4] and usually occurs in the form of agglomerated flakes on the surface as shown in Figure 3. The second depth range, Area B, represents the near-surface contamination layer where the aluminium content decreases linearly with increasing number of laser pulses or depth, respectively. Area C finally shows the bulk material without any aluminium contamination. This is found after five laser pulses at the centre and not reached even after twelve laser pulses at the edge where the contamination layer is thus much thicker.

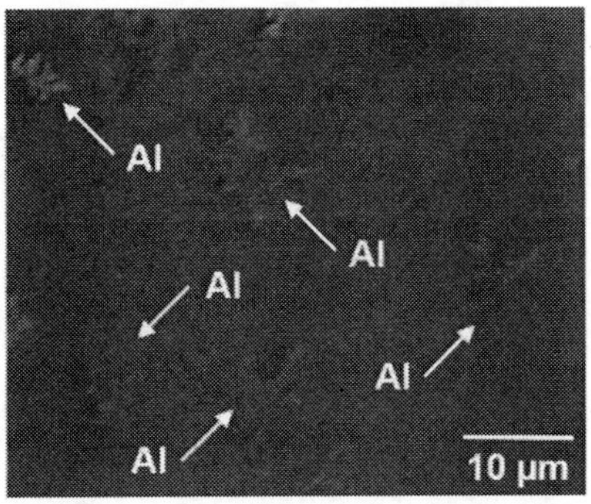

Fig. 3. Scanning-electron microscopy image of a classically-manufactured fused silica surface, taken at the edge of the component. The whitish flakes are aluminium-containing compounds as verified by energy-dispersive X-ray spectroscopy.

4 Conclusions

The detected local contaminations and the accompanying non-uniformity in chemical composition of fused silica windows has a notable impact on the uniformity of the optical properties. As verified via ellipsometry, the index of refraction features an inhomogeneity of about 1.15% over the surface. This effect has severe consequences on the performance of subsequently applied optical coatings. The presented results are thus of notable interest for the manufacture of coated optical components.

This project has received funding from the European Union's Horizon 2020 research and innovation programme under grant agreement no 654148 Laserlab-Europe. The authors further thank Daniel Tasche from the University of Applied Sciences and Arts in Göttingen and Olivier Uteza from the Laboratoire Lasers, Plasmas et Procédés Photoniques in Marseille for their assistance.

References

1. C. Gerhard, D. Tasche, O. Uteza, J. Hermann, Appl. Opt. **56**, 7427 (2017)
2. C. Gerhard, J. Hermann, L. Mercadier, L. Loewenthal, E. Axente, C.R. Luculescu, T. Sarnet, M. Sentis, W. Viöl, Spectrochim. Acta B **101**, 32 (2014)
3. C. Gerhard, *Optics Manufacturing: Components and Systems* (CRC Taylor & Francis, Boca Raton, 2017)
4. C. Gerhard, T. Weihs, D. Tasche, S. Brückner, S. Wieneke, W. Viöl, Plasma Chem. Plasma P. **33**, 895 (2013)

A new approach to laser polishing and form correction of optical components

Harold Kessler[1,*], *Roelene* Botha[1], and *Carsten* Ziolek[1]

[1]NTB Interstaatliche Hochschule für Technik Buchs, Institute for Production metrology, Materials and Optics, 9471 Buchs SG, Switzerland

Abstract. The increasing variety of optical components and materials, combined with stricter surface tolerance requirements, necessitate refining existing polishing processes and developing innovative new polishing solutions and metrology technologies. A fast, reproducible laser polishing process would offer considerable economic benefits over conventional mechanical polishing processes and interest a broad variety of optics manufacturers. In this work, a holistic approach is taken to address the various aspects of glass polishing and form correction via a novel laser polishing system design, the use of a measurement strategy that can be integrated inline and simulation results that are correlated with process parameter studies for different materials.

1 Introduction

The use of aspheres have been attracting a large amount of attention as they enable lens system designers to decrease the number of lenses, and consequently the number of optical surfaces, used. This lowered complexity can improve the performance of devices, reduce waste and lower the manufacturing cost. Achieving surface figure and roughness values for such freeform optics that are comparable with those of standard precision optical components still remains elusive and requires a full understanding of the complete production technology chain. This is not straightforward, when considering the complexity of the current optical component fabrication process. Typical manufacturing steps include shaping, grinding and edge grind and polish processes, followed by iterative intermediate, final and small tool (figure correcting) polishing steps that include intermittent ex-situ surface characterization measurements. The above described conventional methods for polishing optics can rightly be considered as «more an art than a science». [1] The variety of process parameters affect the final output of the process, but their individual and mutual effect is not clear. [2] It is of interest to have a stable and fully understood process independent of material or product shape.

1.1 Laser polishing of glass: State-of-the-art

The main advantage of laser polishing is that it allows avoiding the restrictions of conventional mechanical polishing methods and their tools. Furthermore, polishing errors, such as sub-surface damage, can be completely avoided by using laser polishing.

Laser polishing of fused silica surfaces was first done to increase the damage resistance against laser radiation of shorter wavelengths. [3] Efficient polishing processes of fused silica surfaces down to few nanometres have been shown by various research teams. [4-6] Hildebrand et al.[4] performed a numerical simulation describing the influence of several process parameters. To monitor the laser polishing and shape correction process, a pyrometer or spectrometer can be used to continuously measure the temperature in the polishing spot and accordingly adjust the laser power [7,8]. The process can also be visualized using imaging techniques based camera systems or laser beams and zoom-telescopes [9,10]. By monitoring the actual laser power using a photo diode in the laser beam delivery system, the laser power can be corrected as required using a proportional–integral–derivative (PID) feedback controller [11]. Many crucial aspect of the laser polishing process, process control and necessary metrology are still not fully understood and many open questions remain.

2 Approach

2.1 Considerations for the laser polishing strategy

This work aims to establish a fully understood, reliable and automated laser polishing process to produce freeform optics with surface figure and roughness values comparable with those of precision optical components. A simulation considering additional temperature dependent effects to describe the finished surface morphology is also developed. The following points are taken into consideration for the development:

• Selection of the polishing strategy.

* Corresponding author: harold.kessler@ntb.ch

The polishing strategy between different research teams is strongly varying. Some use fixed parameters while others regulate their process with feedback loops. The impact of different polishing strategies on the surface, as well as trade-off between expense and achieved result will be analysed.

• Increasing the variety of materials that can be processed.

Fused silica is the material of choice for laser polishing, especially with CO_2 laser polishing. This is due to the ideal absorption characteristics of fused silica in this wavelength regime. Weingarten et al. [6] further successfully polished BK7 and S TIH6 in their work.

• Correlating the simulation results with the observed material changes.

Being able to simulate the process for different parameters and materials is desirable. An approach for the simulation was already done by Hildebrand et al. [4]. This model is analyzed and further extended with additional parameters.

• Developing a surface measurement strategy that can be implemented in-line.

In-situ measurement of the samples during the polishing process can be used to further reduce the time needed compared to conventional polishing.

• High temperature stability of the polished components.

Some of the physical effects utilised for laser polishing, e.g. for the form correction after polishing can be reversed with annealing. This effect has to be taken into account.

2.2 In-line measurement strategy

The measurement of the roughness and form of the optical component needs to adapt to the component's shape. As free-form optics are becoming increasingly important, this constitutes a serious obstacle. Optical measuring methods such as chromatic confocal microscopy, laser scanning microscopy, Phase Measuring Deflectometry (PMD), white light interferometry and optical coherence microscopy, as well as a combination between an optical measuring technology and AFM, may offer the required flexibility and resolution. Furthermore, acquisition of measurement data in real-time could also be used to improve the stability of the polishing process.

In the PMD method [12] shown in Fig. 1, two-dimensional sinusoidal fringes are projected onto the surface. Varying distances between the projector and the surface will result in an observable change of the fringes. This variation is recorded by a camera. The fringes are phase shifted and another picture is taken. After several iterations, the surface profile and curvature can be deduced. PMD is a rather cost effective and robust way to measure 3D freeform surfaces, especially for specular reflection surfaces.

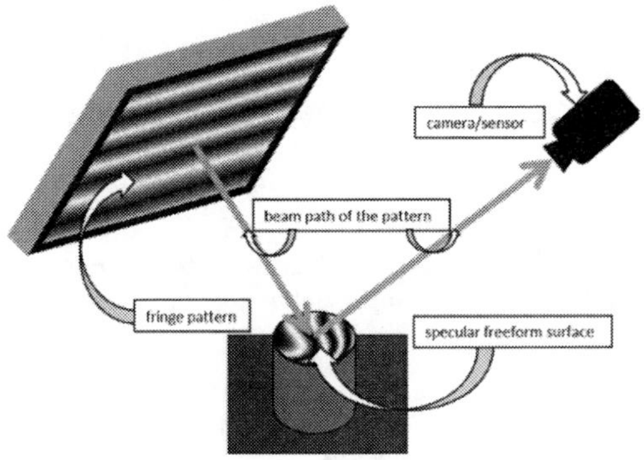

Fig. 1. Schematic of the PMD setup.

3 Conclusion

This work discussed the considerations that were taken into account for developing a new laser polishing process for optical freeforms. A concept for a CO_2 laser polishing system was discussed and a surface measurement strategy that can be implemented in-line suggested. We have started studying the process windows for fused silica and sapphire. In a consequent step, the results of the parameter study will be compared with simulation results. A comparison of the influence different polishing strategies have on the efficiency and the stability of the polishing process will then be performed.

References

1. I. D. Marinescu, E. Uhlmann, T. Doi, *Handbook of Lapping and Polishing*(CRC Press, 2006)
2. D.K. Mandal, C. S. Syan, *CAD/CAM, Robotics and Factories of the Future*)
3. P.A. Temple, W. H. Lowdermilk, D. Milam, Appl. Opt. **12** (1982)
4. J. Hildebrand, K. Hecht, J. Bliedtner, H. Müller, Physic Procedia, **39**, 277-285 (2012)
5. M. J. Matthews, S. T. Yang, N. Shen, S. Elhadj, R. N. Raman, G. Guss, I. L. Bass, M. C. Mostrand, P. J. Wegner, Adv. Eng. Materials **17**, 247-252 (2015)
6. C. Weingarten, A. Schmickler, E. Willenborg, K. Wissenbach, R. Poprawe, J. Laser Apps. **29** (2017)
7. S. Heidrich, A. Richmann, P. Schmitz, E. Willenborg, K. Wissenbach, P. Loosen, R. Poprawe, Optics and lasers in Engineering **59**, 34-40 (2014)
8. B. Burzic, M. Hofele, S. Mürdter, H. Riegel, J. of Laser Applications, **29**, 011701 (2017)
9. O. Pütsch, A. Temmler, J. Stollenwerk, E. Willenborg, P. Loosen, Proc. SPIE 8843 (2013)
10. A. Temmler, E. Willenborg, K. Wissenbach, Proc. SPIE **8243**, 82430W (2012)
11. E.V. Bordatchev, 3rd LAP Conf, Aachen, DE (2018)
12. F. Z. Fang, X. D. Zhang, A. Weckenmann, G. X. Zhang, and C. Evans, CIRP Annals, **62** 823–846 (2013)

 https://doi.org/10.1051/epjconf/201921509002

Computer Generated Holograms (CGH) for Testing, Alignment and Positioning of Astronomical and Space Mirrors Aspherical Surfaces

Aleksandr Semenov[1,*], *Magomed* Abdulkadyrov[1], *Nikolay* Dobrikov[1], *Aleksandr* Ignatov[1], *Vladimir* Patrikeev[1], *Vitaliy* Pridnya[1], *Andrey* Polyanchikov[1], *Ruslan* Nasyrov[2]

[1]JSC Lytkarino Optical Glass Factory (LZOS), Lytkarino, 140080, Parkovaya 1, Moscow region, Russia
[2]Institute of Automation and Electrometry of the Russian Academy of Science, 1 Koptyuga ave. Novosibirsk, 630090, Russia

Abstract. JSC LZOS successfully applies Computer Generated Holograms (CGH) for testing of on- and off-axis surfaces of large-dimensioned optical components for their alignment with respect to interferometer, for defining of aspherical surface optical vertex position with respect to the geometrical center and for optical systems alignment. All this allows producing of large-dimensioned aspherical on- and off-axis astronomical and space mirrors with high accuracy.

1 Introduction

JSC LZOS has been using Computer Generated Holograms (CGH) [1-3] for testing of surfaces of large-dimensioned optical components for astronomical and space application for many years. They are applied in the process of testing of aspherical surface figure, testing of off-axis aspherical surfaces with fixing of off-axis parameters and mirror orientation, testing of aspherical surface vertex position with respect to the mirror geometrical center, mutual alignment of mirrors in test scheme, etc. Thus, CGHs have become an integral part of up-to-date testing of aspherical surfaces of large-dimensioned optical mirrors and optical systems.

2 Specific Features of Aspherical Surfaces Testing by means of CGH

Let us consider aspherization of off-axis mirror of diameter 560 mm and asphericity 200 μm with distance from the component center to the optical axis about 480 mm. Final polishing was carried out with testing in scheme including CGH [4,5]. Fig. 1 shows testing scheme including CGH corrector with aligning focusing elements (Fig. 2) for CGH alignment with respect to interferometer. Focusing elements create specular points, according to which the component alignment with respect to the CGH corrector is performed, in the edge zone of the component, in the center and in the vertex position of the optical axis (Fig. 1).

Accurate and unambiguous mirror alignment in space allowed successful performance of operations on figuring of aspherical surface. The obtained quality of hyperboloid deviations from theoretical RMS=0.05λ (λ=632.8 nm).

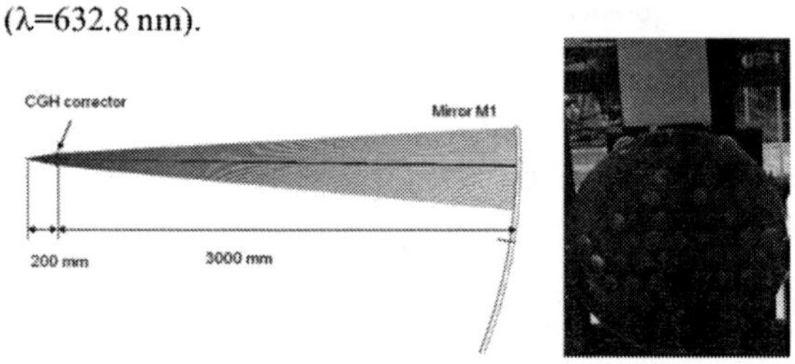

Fig 1. Testing scheme and focusing alignment elements of off-axis mirror of diameter 560 mm

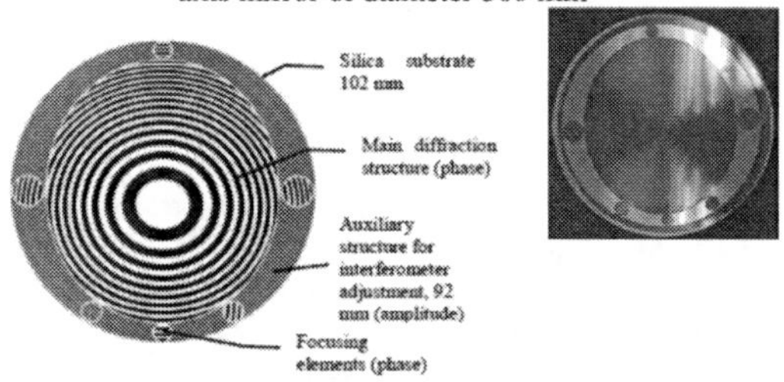

Fig. 2. CGH for off-axis mirror testing.

Another example of testing of of-axis optical elements is shown in Fig. 3. It shows testing of stereoscopic system with two off-axis mirrors, positioned symmetrically with respect to the line crossing the optical axis perpendicularly. The mirrors are installed on Whiffle tree. Then one mirror unit with support is fixed motionless, while the second one (or both) is installed on hexapod. Thus, mutual alignment of off-axis elements by means of hexapod and CGH located in the mirrors testing point is performed. CGH for

* Corresponding author: lastro@lzos.ru

mirrors testing is shown in Fig. 3. It has special auxiliary diffraction structures for the formation of crosses in the mirror area, which are used for alignment of mirrors with respect to CGH.

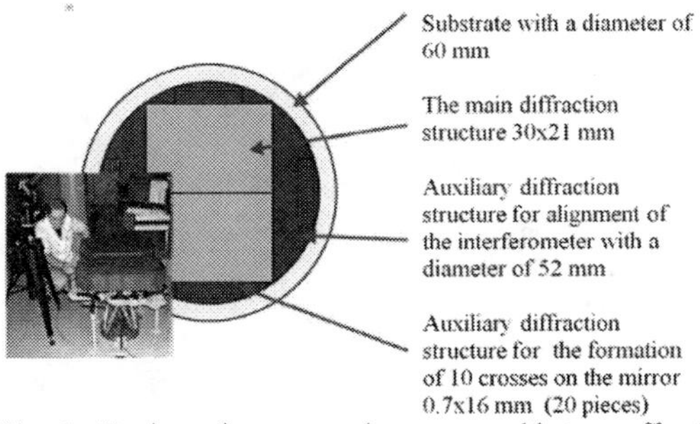

Fig. 3. Tetsing of stereoscopic system with two off-axis mirrors.

Under Contract with AMOS, JSC LZOS is producing optics for DAG Telescope (Doğu Anadolu Gözlemevi) [6]. Primary mirror DAG M1 of diameter 4000 mm has asphericity 139 μm and aperture f/1.8 and very strict requirements to the accuracy of optical surface production. CGH corrector was designed and produced by the Institute of Automation and Electrometry of the Russian Academy of Science Siberian Department (Novosibirsk). It converts spherical wavefront into aspherical and back. The corrector was produced from quartz plate of diameter 102 mm, corrector diameter is 80 mm, the diameters of the auxiliary adjustment sphere circle are 82-92 mm. The distance from the hologram to the focus point is 1170 mm. In addition, this wavefront corrector has one more circular structure focusing the hologram, which creates a specular point in the mirror vertex (Fig. 4). By means of this point, we are able to define the aspherical surface vertex shift with respect to the mirror geometrical center, which is required by the specification. The accuracy of positioning of such point depends on the accuracy of CGH installation with respect to the interferometer. When CGH works from diverging beam, it is adjusted by means of reflecting hologram. Such reflecting hologram is calculated in such a way that an endless strip appears on it when CGH is accurately aligned with respect to the interferometer. Focusing structure for DAG M1 was made in the form of circle around the main structure. Such structure forms a point in the mirror vertex. Distance Focus-CGH = 1170 mm, Distance CGH-Mirror = 14274 mm, Retro Diameter (Out-Inner) = 86-96 mm, Focusing Diameter (Out-Inner) = 80.4-84.4 mm.

Fig. 5 shows scheme of the focused point shift when CGH is tilted. It can be seen from the Figure that when CGH is tilted by angle α, the point shift is d=L*tan(α). Obviously, the hologram tilt when there is one strip on the reflecting structure does not depend on distance расстояния F, it is defined by the diameter of the reflecting hologram only. The spot size is defined on the basis of Rayleigh criterion 2.44 λL/D, where D is diameter of the focusing structure and λ= 0.6328 μm is the interferometer wavelength. In is apparent that the vertex shift (48 μm) is less than the spot size (260 μm). However, in practice, it is difficult to set even one strip, and there are usually three-four of them. Therefore, it is necessary to multiply the vertex shift by the number of strips. This is true for a small number of strips on interferogram.

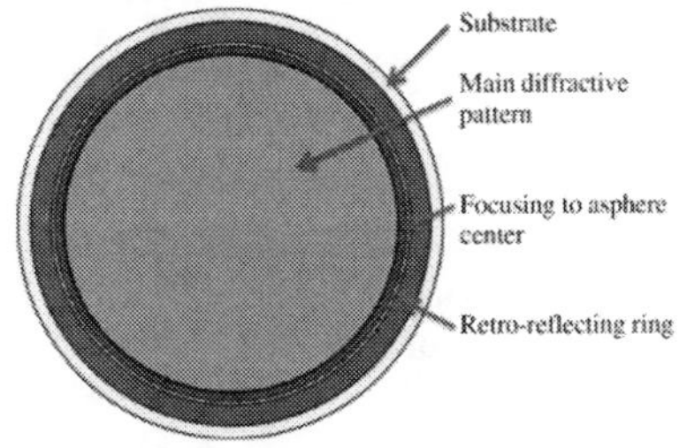

Fig. 4. Hologram layout.

Omitting a more detailed description, we can conclude that the indicated method allows defining of the mirror vertex position with accuracy about 100 μm.

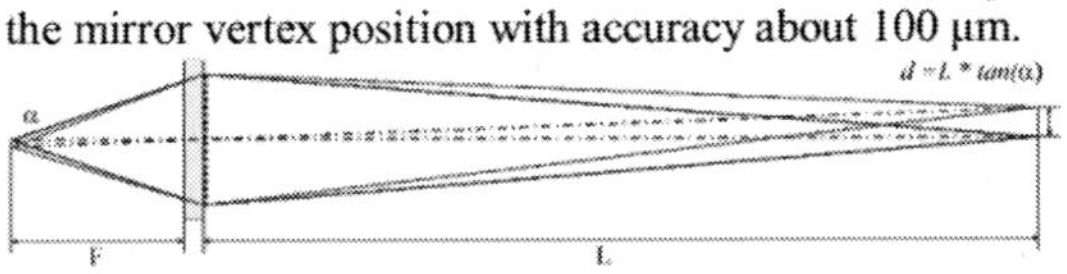

Fig. 5. Shift of the surface vertex marked point due to hologram tilt with respect to interferometer.

References

1. A.G. Poleshchuk, V. P Korolkov, R. K. Nasyrov, J.M. Asfour, Proc SPIE **7102**, 710206 (2008)
2. A.G Poleshchuk, Proc. SPIE **7544**, 75443L-1 (2010)
3. A.G. Poleshchuk, Proc. SPIE **7133**, 713333-1 (2009)
4. A.P. Semenov, M.A. Abdulkadyrov, V.E Patrikeev, Proc. SPIE 9683, 968314-1 (2016)
5. A.P. Semenov, M.A. Abdulkadyrov, V.E. Patrikeev, Proc. SPIE 9912, 991246-1 (2016)
6. M. A. Abdulkadyrov, A. P. Semenov, S. P. Belousov, N. M. Vladimirov, A. N. Ignatov, V. E. Patrikeev, V. V. Pridnya, A. V. Polyanchikov, R. K. Nasyrov, O. Pirnay, C. Flebus, C. Yesilyaprakd, O. Keskine, Proc. SPIE 10706, 1070631-1 (2018)

New Technical Solutions in the Production Method of High-Aperture Ruled Diffraction Gratings

Andrei Melnikov*, and *Anatoly* Lukin

JSC "Scientific and Production Association "State Institute of Applied Optics", Civilian Production Department, 2, N. Lipatova st., Kazan, Russia, 420075

Abstract. This paper presents a new approach in the production method of high-aperture diffraction gratings with high diffraction efficiency on convex and concave working surfaces with large deflection based on the use of an original pendulum-type ruling engine.

1 Relevance

A major optical element of a wide range of spectral equipment is a nonplanar diffraction grating with high diffraction efficiency and high aperture which, besides its primary function as a dispersive element, is involved in correcting aberrations and imaging the objects under study.

The use of these diffraction gratings is especially relevant for developing compact high-aperture spectrophotometric devices for remote probing of the Earth from the outer space, ecological monitoring, etc. Moreover, it opens up the possibility to minimize the number of optical elements in the layout of such equipment.

Two main techniques are used today to make high-aperture gratings: ruling method (based on the classic Rowland ruling engines and precision lathes with numerical program control) and holographic method [1-6]. However, these techniques do not allow making diffraction structures on nonplanar working surfaces with large deflection (5 mm and more) while keeping a high concentration of energy in the diffraction working order and a moderate level of scattered light [7-10].

2 A pendulum-type ruling engine

A new technical solution suggested by us enables to overcome the aforementioned problem in the production method of high-aperture ruled diffraction gratings [11]. Figure 1 below represents a general view of a pendulum-type ruling engine for making grooved structures of ruled diffraction gratings on convex spherical or aspherical (including toroidal) working surfaces.

The blank carriage is set on a base turnable on rotation bearings in a plane perpendicular to the plane of angular auto-oscillations of the diamond carriage; the geometrical pivot axis of the blank carriage intersects the geometrical axis of angular auto-oscillations of the diamond carriage. The substrate of the manufactured convex diffraction grating is attached to the blank carriage; a diamond cutter with a hoisting and lowering mechanism is attached to the diamond carriage. The blank carriage drive is made with the function of electronic control of its turning process.

The diamond carriage is a physical pendulum, the angular auto-oscillation axis of which is secured on two rotation bearings that ensure large angular auto-oscillation amplitude in the swinging plane. This carriage is made as a vertical frame and equipped with a telescopic mechanism for height adjustment. Due to having a counterweight, the diamond carriage possesses the dynamic properties of a physical pendulum with a center of mass situated below the geometrical axis of its angular auto-oscillations. The movement trajectory of the diamond cutter during its cutting pass is almost equal to the curvature radius of the convex working surface of the substrate. The diamond carriage drive is intended to provide lateral motion, i.e. diamond cutter feed in the form of continuous angular oscillations, which is achieved in the auto-oscillation mode.

Unlike other known analogues, the suggested construction principles of this ruling engine enable to form grooved structures on nonplanar working surfaces almost with no deflection limitations.

The version of ruling engine construction shown in Figure 1 enables to produce diffraction gratings only on convex surfaces. To get high-aperture diffraction gratings on concave working surfaces, it is possible to use the technique of precision replication with the use of low-shrinkage polymeric compositions [12]. In this case, a convex original grating made with the suggested ruling engine can be used as a master matrix.

This two-stage approach to grating production ensures a rather high optical quality and high diffraction efficiency, as well as low cost and high productivity, which is especially important in the context of large-scale production of spectral equipment of various applications.

* Corresponding author: mr.melnikov@bk.ru

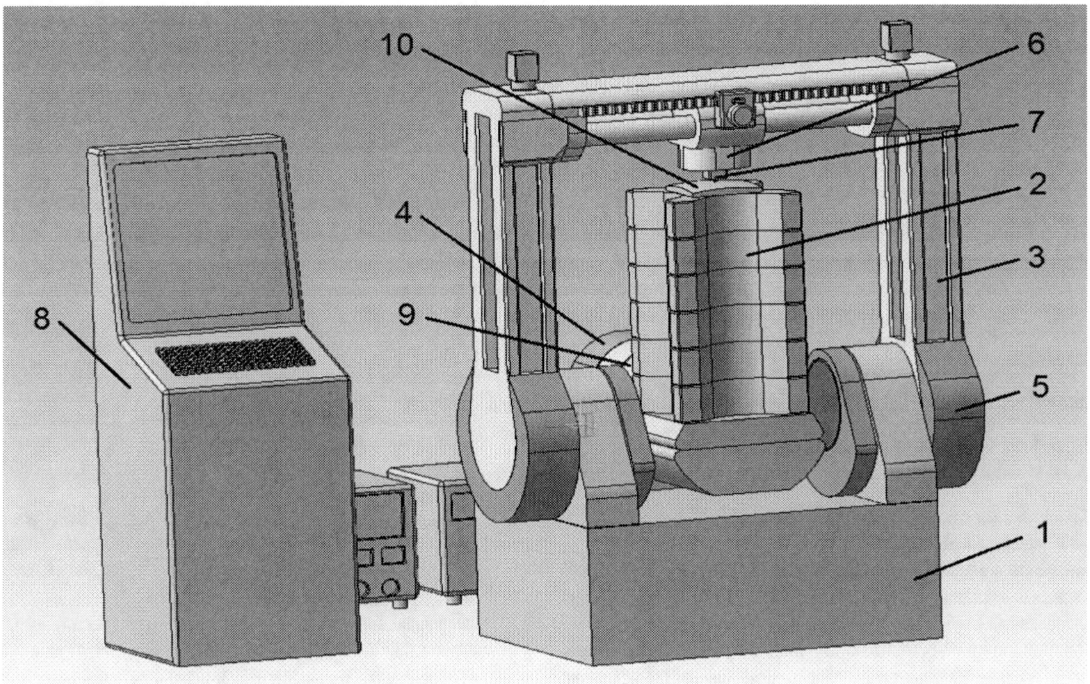

Fig. 1. The main units and elements of a pendulum-type ruling engine: 1 – base; 2 – blank carriage; 3 – diamond carriage; 4 – blank carriage drive; 5 – diamond carriage drive; 6 – diamond cutter hoisting and lowering mechanism; 7 – diamond cutter; 8 – blank carriage drive, diamond carriage drive, and diamond cutter hoisting/lowering mechanism control unit; 9 – angular position sensor of the blank carriage; 10 – substrate of the ruled diffraction grating that is being manufactured.

3 Conclusion

In view of this, the new suggested technical solutions open the potential to produce high-aperture diffraction gratings with high diffraction efficiency on convex and concave working surfaces for compact spectral equipment of various applications.

References

1. M.C. Hutley, *Diffraction gratings* (Academic Press, London, 1982)
2. F.M. Gerasimov, E.A. Yakovlev, *Modern trends in the spectroscopy technique* (Nauka, Novosibirsk, 1982) [in Russian]
3. T. Harada, T. Kita, Appl. Opt. **19**, 3987-3993 (1980)
4. C. Palmer, *Diffraction grating handbook* (Newport Corporation, N.Y., 2014)
5. D. Xu, J.D. Owen, J.C. Papa, J. Reimers, T.J. Suleski, J.R. Troutman, M.A. Davies, K.P. Thompson, J.P. Rolland, Opt. Exp. **25**, 15252-15267 (2017)
6. T. Sukegawa, Yu. Okura, T. Nakayasu, *Proc. SPIE* **10706**, 107063L-1-107063L-6 (2018)
7. JSC Shvabe Holding / JSC Scientific and Production Association “State Institute of Applied Optics” / Optic component base catalogue / URL: http://shvabe. com
8. Hitachi Group Corporation / URL: http://www.hitachi-hightech.com
9. HORIBA Jobin Yvon Ltd. Corporation / URL: http://www.horiba.com
10. Carl Zeiss Microscopy GmbH Company / URL: http://www.zeiss.de/gratings
11. A.V. Lukin, A.N. Melnikov, Useful model patent RU № 185039 dated 19 Nov 2018 [in Russian]
12. M.M. Akhmetov, A.F. Belozyorov, V.A. Baloyev, A.A. Belokopytov, I.S. Gainutdinov, V.P. Ivanov, A.V. Lukin, A.N. Melnikov, I.A. Mogilyuk, Kontenant **15**, 39-42 (2016) [in Russian]

Fabrication of SiC Off-axis Aspheric Mirror by Using Robot Polishing

Haitao Liu*, *Fengtao* Yan, *Wenchuan* Zhao, *Jieli* Wu, and *Min* Zhou

Institute of Optics and Electronics, Chinese Academy of Sciences, 610209 Chengdu, China

Abstract. Two silicon carbide (SiC) lightweight off-axis aspheric mirrors were fabricated by using robot polishing technology. One is a primary mirror with dimension 380mm×328mm, the other is a tertiary mirror with dimension 294mm×238mm. The robot polishing technology is the combination of CCOS technology and a 6-axises industrial robot. The setup of robot polishing system is discussed, and the mirror aspheric grinding, surface polishing and figuring process are reported in this paper. The final surface shape error of primary mirror is 11.4nm RMS, and the tertiary is 12.1nm RMS. The capability of off-axis aspheric mirror fabricating on robot polishing system is verified.

1 Introduction

SiC has excellent material property such as high thermal and mechanical stability, which is suitable for optical application in aerospace and next generation lithography. Off-axis aspheric mirrors is widely used in many optical systems [1]. For SiC off-axis aspheric mirrors, the fabricating quality, efficiency and costs are three main problems. The Computer Controlled Optical Surfacing (CCOS) technology has long been recognized as a suitable way to grinding and polishing aspheric surface [2, 3].

In this paper, we introduce the setup of the robot polishing system, and present the process of SiC mirror aspheric grinding and polishing. And the fabrication results is shown.

1.1 Robot polishing system

We use a Stäubli TX200 heavy duty 6-axises industrial robot as the motion platform. The polishing tool is a spin-orbital due-rotate sub-aperture lap which mounted on the robot. The polishing force is automatically applied by pneumatic cylinder in order to obtain stable force load during the polishing.

Fig. 1. Setup of robot polishing system.

In order to make the robot suitable CCOS fabrication, we developed a software in VAL3 language which runs on the robot controller in order to interpret the G-code like polishing path file to the robot moving code.

The specifications of robot polishing machine are list below.

- Maximum work range: 2 m
- Trajectory accuracy: ±0.1 mm
- Max. tool spin speed: 300 rpm
- Max. tool orbital speed: 200 rpm
- Max. polishing force load: 30 N
- Capability: plane, sphere, asphere and freeform surface's grinding and polishing

1.2 Specification of the SiC mirrors

The primary and tertiary mirror are both high order off-axis aspheric mirrors with rectangle shape. The dimension of primary mirror is 380mm×328mm, and the tertiary mirror is 294mm×238mm. The material is reaction burned silicon carbide (RB-SiC). The requirement of surface shape error is λ/50 RMS for these two mirrors (λ=632.8nm).

The initial surface is best fitting sphere, so the aspheric deviation is quite large. The initial shape error of primary mirror is 0.2mm PV, and tertiary mirror is 0.08mm PV. So before the polishing, the aspheric grinding is used in order the remove the large surface shape error.

2 Aspheric grinding of SiC mirrors

In grinding, the copper pad is used in order to remove the bulk errors, tool marks and sub-surface damages. The pad size is from 15mm to 50mm in diameter, determined by the spatial frequency of the surface shape error that need to be removed. The diamond abrasives with different particle size are used in different phase. There are 4 size abrasives, which are W40, W20, W10 and W5, has been used in grinding.

* Corresponding author: lhtioe@163.com

The surface shape is measured by LEITZ ultra CMM in the grinding. We developed a program to analysis the error map by using the multi dimension optimize algorithm. The surface shape error can be reduced to about 2um PV after grinding process, see figure 2.

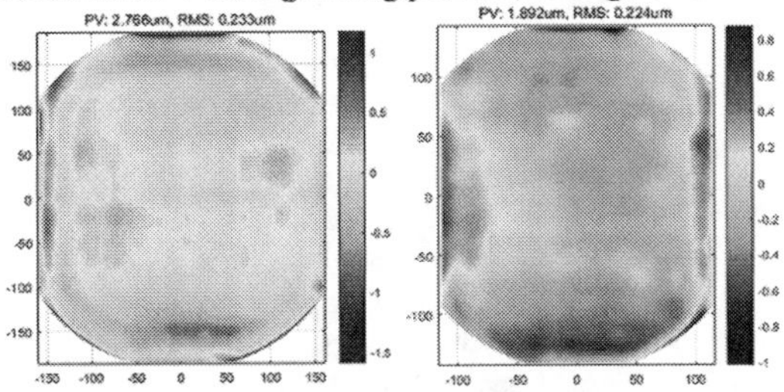

Fig. 2. Surface shape error after grinding. (Left: primary mirror, PV 2.8um. Right: tertiary mirror, PV 1.9um)

3 Aspheric polishing of SiC mirrors

Polishing process includes SiC substrate polishing before surface modification, Si substrate polishing after surface modification, and surface figuring. Surface modification process is an efficient way that can improving SiC surface compactness by depositing a thin layer of Si in the surface.

In SiC substrate polishing, the material of polishing pad is polyurethane, and the abrasive is W2 diamond polishing slurry. The main goal is to polish the surface to make it suitable for surface modification, and reduce surface shape error. By using the robot polishing, the surface shape error can be reduced to $\lambda/10 \sim \lambda/20$ RMS, and the surface roughness can be reduced to less than 10nm Ra. Figure 3 shows the polishing of primary mirror.

Fig. 3. Primary mirror substrate polishing by robot.

Ion beam assisted deposition method is used in surface modification. There is about 10μm thickness Si layer deposited in the SiC substrate surface. The surface shape error is almost maintained the same after the surface modification, but the roughness is increased due to the deposition. So the mirror surface need to be re-polishing.

The final surface figuring is completed by combine the robot polishing and magnetortheological finishing (MRF). The MRF is used to achieve high accuracy figuring and the robot polishing is used to smoothing the surface to remove middle-high spatial frequency shape error. The finished mirror surface is shown in figure 4.

During polishing, the aspheric surface shape error is measured through computer generated holograph (CGH) null-compensation interferometry. A 6 inch Zygo Verifire interferometer with 1K×1K pixels CCD is used in the metrology.

Fig. 4. The finished mirror surface. (Primary mirror)

4 Results and discussion

Through several iteration of MRF figuring and robot polishing, the surface shape error can be reduced to satisfied the requirement. The surface shape error for primary mirror is 11.4nm RMS (see figure 5), and for tertiary mirror is 12.1nm RMS (not show here).

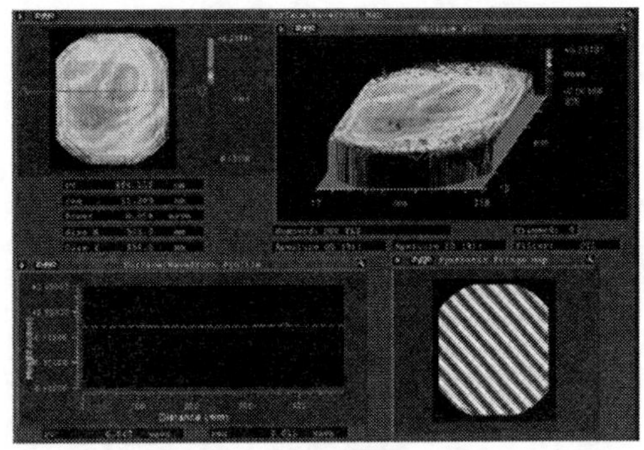

Fig. 5. Final surface shape error of primary mirror.

Surface roughness is also measured by using Bruker white light interferometer, and the roughness is about 1.5nm~3nm in Ra. Figure 6 show one of the results of primary mirror.

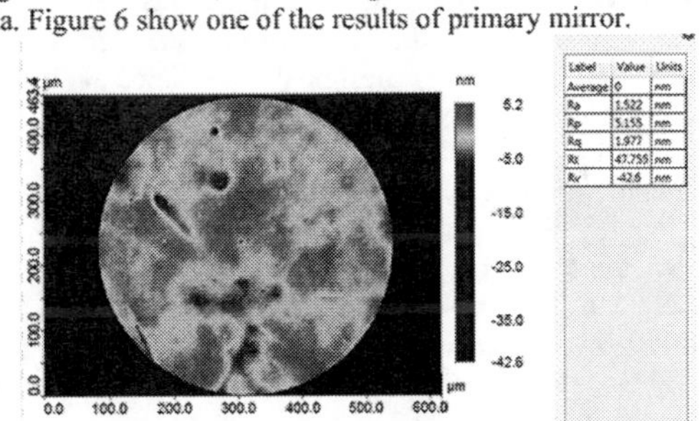

Fig. 6. Surface roughness of primary mirror.

References

1. C. Atkinson, S. Texter, Status of the JWST optical telescope element, Proc. of SPIE, Vol. 8442, 2012.
2. R. A. Jones, Fabrication of off-axis optical segments, Proc. of SPIE, vol. 1752, 1992.
3. D. D. Walker, C. Dunn, G. Yu, et. al., The role of robotics in computer controlled polishing of large and small optics, Proc. of SPIE, vol. 9575, 95750B, 2015.

https://doi.org/10.1051/epjconf/201921509005

Fast Fabrication of 1.26-Meter-Long Optic for Elliptic Cylindrical Focusing Bending Mirror

Jieli Wu[1,*], *Haitao Liu*[1], *Fengtao Yan*[1], *Min Zhou*[1], *Bing Zhuo*[1], *Wan Yongjian*[1]

[1]Institute of Optics and Electronics, Chinese Academy of Sciences, NO.1 Guangdian Avenue Chengdu, China

Abstract. As beamline emittance reaching the diffraction limit, the demand for focusing mirrors over one meter in length is ever increasing. Bending mirrors, are competitive for its adjustable radius and high manufacture precision, whose final surface form is managed by a delicate bending of a high precision plane. A special contour plays an essential role and benefits this final touch, given that the computed torque along the length closely correlates with the corresponding width of the optics. However, it poses main challenges with severe edge effect, deformation and metrology to the fabrication of the precursor plane mirror. This paper reports on a fast processing technology for a 1.26-meter-long plane as a bending mirror with irregular shape. A smooth surface with a slope error of 0.67 urad(RMS) and 0.3nm roughness is achieved through deterministic flexible finishing of robot polishing system and annular polishing, providing a high precision and cost-effective method to meet the growing needs for long X-ray mirrors.

1 Introduction

Highly brilliant and coherent X-ray beams are provided by undulators at the third-generation synchrotron radiation sources such as Advanced Photon Source (APS), European Synchrotron Facility (ESRF), SPring-8 and Shanghai Synchrotron Radiation Facility (SSFR) [1, 2]. As the development of synchrotron radiation sources, the demand for focusing mirror over one meter in length is greater than ever, which brings bigger challenge to figuring.

Bending mirror, with the advantage of adjustable radius and relative high manufacture precision, is more applicable for elliptic cylindrical mirrors of this length. Yet, in order to bend the plane mirror into ideal surface form, the contour is designed into special shaped according to the computed torque. Edge effect, deformation and sub-aperture stitching measure error inevitably hinders the surface precision optimization for the lengthy and special- shaped optics.

In this research, deterministic flexible finishing of robot polishing system and annular polishing were utilized in sequence which realized a high figuring efficiency and smooth surface for the 1.26-meter-long heteroideus plane.

2 Fabrication system and surface figuring method

2.1 1.26-Meter-long heteroideus plane

The parameter of the 1.26-meter-long heteroideus plane is listed in the Table 1. The shape is composed of segmental straight lines along the longitude direction which were symmetrical in width direction as shown in figure 1.

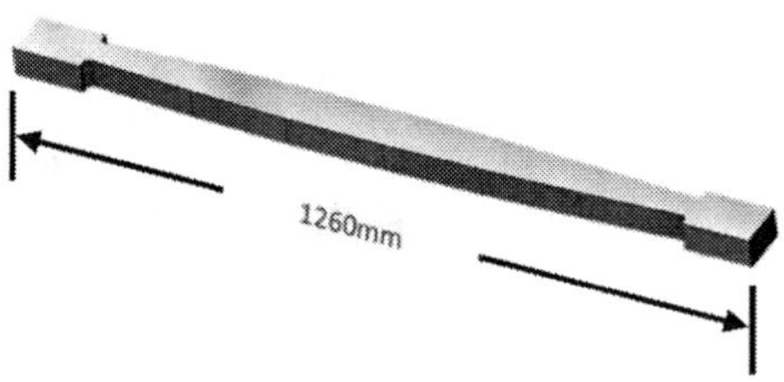

Fig. 1. 1.26-meter-long heteroideus plane

Table 1. Formatting sections, subsections and subsubsections.

Substrate material	(110)Si single crystal
Mirror size	1260mm*80mm*50
Effective mirror size in longitude direction	1000mm
Acceptance width	20mm

The figuring of the optic includes finishing of robot system and smoothing of annular lapping which processed sequentially.

2.2 Deterministic flexible finishing of robot polishing system

* Corresponding author: author@e-mail.org

https://doi.org/10.1051/epjconf/201921509005

Like the conventional polishing, material removal of robot polishing system is based on the Preston equation, yet the flexible 6-axis moving arm and the 2-meter arm reach of robot polishing system make it a cost-effective finishing method for large unconventional optics. The typical procedure for the heteroideus plate are shown in figure 2, in which the edge effect suppression is added. The edge warp was successfully restrained below 2um and smoothly connected to the near area.

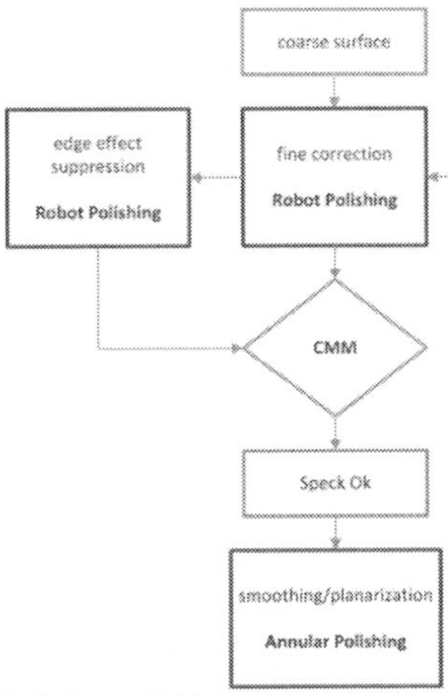

Fig. 2. Robot polishing process

Fig. 3. Robot polishing

2.3 Annular Polishing

Further figuring was performed on a 5m annular polishing machine shown in figure 3. With the optimized pitch grid size, fine abrasive, slurry concentration and velocity, the plane was further planarized and smoothed.

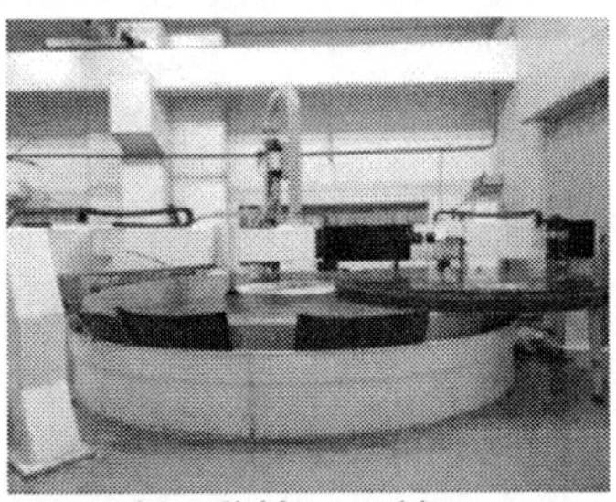

Fig. 4. 5-meter annular polishing machine

Since the polishing lap of annular polishing is much larger than the optic, it offers a global surface planarization requiring no stitching measurement during the process. The plane was then fast figured and smoothed.

3 Results and discussion

The fabricated surface accuracy was measured by Zygo Verifire 6'' interferometer along the longitude direction, with 11 measuring spots to cover the full-aperture surface. The slope errors of the 11 spots have preferable consistency around 0.67urad due to the planarization of annular polishing (fig.5 and fig.6). The surface roughness was measured to be 0.297nm (RMS) by NanoCam Sq Dynamic Optical Profiler (fig.7).

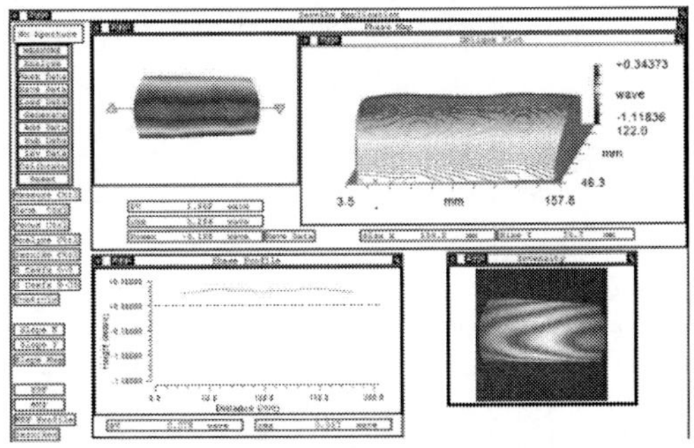

Fig. 5. Typical surface form accuracy of the fabricated mirror

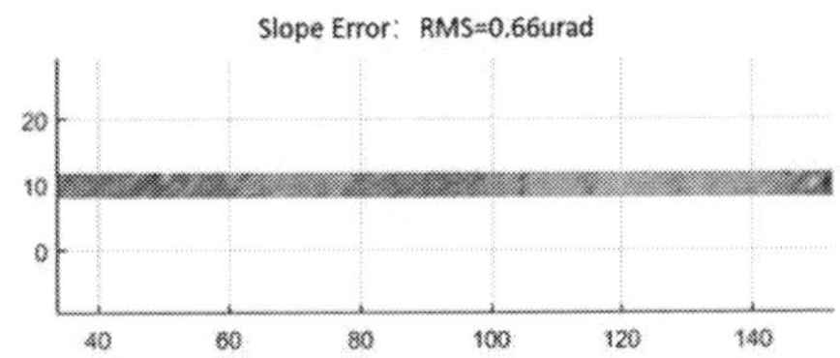

Fig. 6. Typical surface form accuracy of the fabricated mirror

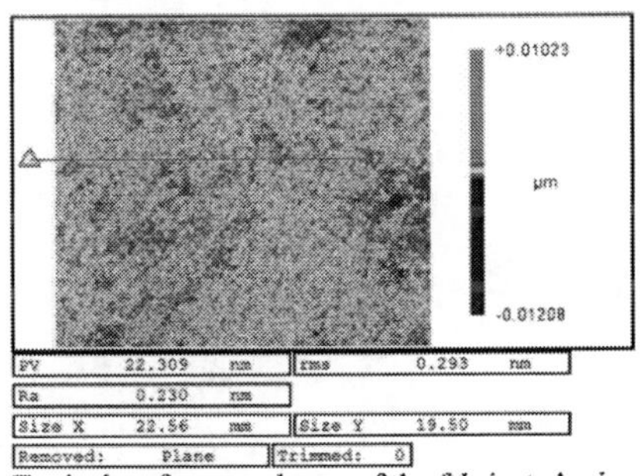

Fig. 7. Typical surface roughness of the fabricated mirror

References

1. S. Mobilio, F. Boscherini, C. Meneghini, *synchrotron radiation: Basics, methods and applications* (Springer-Verlag Berlin Heidelberg, 2015)
2. K. Yamauchi, K. Yamamura; H. Mimura, Y. Sano,; S. Matsuyama, H. Yumoto, K. Ueno, M. Shibahara, K. Endo, M. Yabashi, K. Tamasaku, Y. Nishino, T. Ishikawa, Y. Mori, Proc SPIE, 5533 :116 (2004)

 https://doi.org/10.1051/epjconf/201921510001

Ultra-High-Throughput Multi-Parametric Imaging Flow Cytometry

Andrew deMello*, *Anand* Rane, *Gregor* Holzner, and *Stavros* Stavrakis

Department of Chemistry & Applied Biosciences, ETH Zürich, Vladimir Prelog Weg 1, 8093 Zürich, Switzerland

Abstract. I will present a microfluidic imaging flow cytometer incorporating stroboscopic illumination, for blur-free cellular analysis at throughputs exceeding 100,000 cells per second. By combining passive (inertial or viscoelastic) focusing of cells in parallel microchannels with stroboscopic illumination, such chip-based cytometers are able to extract multi-colour fluorescence and bright-field images of single cells moving at high linear velocities. This in turn allows accurate sizing of individual cells, intracellular localization and analysis of heterogeneous cell suspensions. The method is showcased through the rapid enumeration of apoptotic cells, high-throughput discrimination cell cycle phases and localization of p-bodies.

1 Introduction

Flow cytometry, based on point-based detection schemes, is recognized to be the gold-standard tool for high-throughput manipulation, processing and analysis of cells in flow, but is typically limited in regard to the number of cells that can be interrogated per unit time and the information yield per experiment. To address such limitations, much recent activity has focused on the development of imaging flow cytometry platforms[1]. Imaging flow cytometry combines the high-throughput capabilities of conventional flow cytometry with single-cell imaging and enables high information content analysis of large cellular populations. We have developed approaches for sheathless imaging flow cytometry that incorporate stroboscopic illumination for blur-free cellular analysis at throughputs exceeding 100,000 cells per second[2]. By combining inertial (or viscoelastic) focusing of cells in parallel (or single) microchannels and stroboscopic illumination, such chip-based cytometers are able to extract multi-colour fluorescence, bright-field, and dark-field images. This enables the accurate sizing of individual cells and the analysis of heterogeneous cell suspensions while maintaining operational simplicity..

2 Method

A schematic of one of microfluidic imaging flow cytometers is shown in Figure 1. In simple terms, the cytometer consists of three sub-components: an input module (R1) incorporating a series of posts to remove cellular aggregates prior to analysis, a cell-focusing manifold (R2) containing six parallel and winding channels (each with a width and depth of 40 and 22 mm, respectively), and an imaging region (R3) where dual-colour fluorescence, bright-field, and dark-field images of the cell flows can be extracted. In addition to lateral focusing, fluid inertia is used to control the axial position of cells within each microfluidic channel and the inter-cellular spacings. To eliminate motion blur associated with CMOS cameras when imaging rapidly moving cells, we implement a stroboscopic illumination scheme and dual-view detection system, to extract blur-free images. Details of this approach have been reported elsewhere[3].

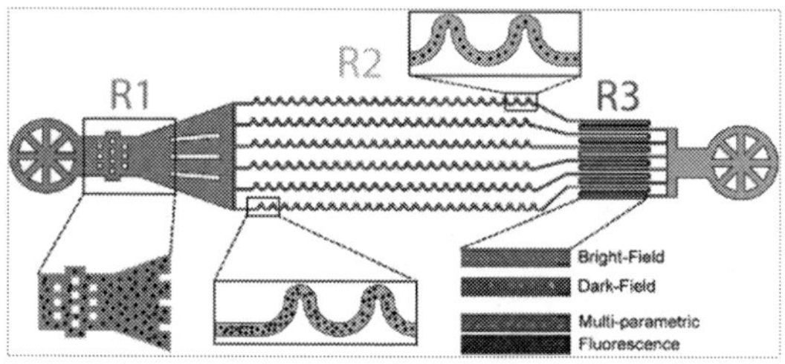

Fig. 1. Schematic of an inertial microfluidic flow cytometer highlighting the inlet port (blue), inertial focusing channels (orange), imaging area for the detection (green), and the outlet port for the collection (gray). (R1) Filter for removing the cell debris, cell aggregates, and the clumps. (R2) Schematic of the cells flowing at the beginning and at the end of the inertial focusing channel. (R3) Representative images from all the possible imaging modalities of the detection. All scale bars represent 30 mm. Image adapted and reproduced from [2].

2 Results

My talk will describe how we have applied the method to the rapid enumeration of apoptotic cells, high-throughput discrimination of cell cycle phases and the localization of p-bodies inside cells[2]. Our approach makes significant contributions to the large-scale screening of cells, allows quantitative analyses of heterogeneous cell populations, and enables the systematic mapping of molecular interactions within real-world cell populations.

References

1. S. Stavrakis, G. Holzner, J. Choo & A.J. deMello, *Current Opinion in Biotechnology* **55**, 36 (2018)
2. A.S. Rane, J. Rutkauskaite, A.J. deMello & S. Stavrakis, *Chem* **3**, 588 (2017)
3. D. Hess, A.S. Rane, A.J. deMello & S. Stavrakis, *Analytical Chemistry* **87**, 4965 (2015)

* Corresponding author: andrew@ethz.ch

 https://doi.org/10.1051/epjconf/201921510002

Learning The Fluid/Flow Properties Using Microfluidics

Pooria Hadikhani[1,*], *Navid Borhani*[1], *S. Mohammad H. Hashemi*[1,2], *and Demetri Psaltis*[1]

[1]Optics Laboratory, School of Engineering, Swiss Federal Institute of Technology Lausanne (EPFL), Lausanne, Switzerland
[2] Computational Science and Engineering Laboratory, ETH Zurich, 8092 Zurich, Switzerland.

Abstract. Deep neural networks (DNN) are employed to measure the flow rate and the concentration of the liquid using the images of the droplets in a microfluidic device. The trained networks are able to measure flow rates and concentrations with good accuracy.

1 Introduction

Flow and fluid properties are crucial parameters that define the droplet flow properties in a microfluidic device. Due to the nonlinear dynamics of fluid flows, it is difficult to extract these properties using the governing equations of fluids. Neural networks can extract the fluid properties from images of their flow patterns since they can deal with nonlinear problems. The schematic of this procedure is shown in Fig. 1. Using the methodology of this study, the trained network is able to identify correctly flow rates and solution concentrations it has not seen before in the training set.

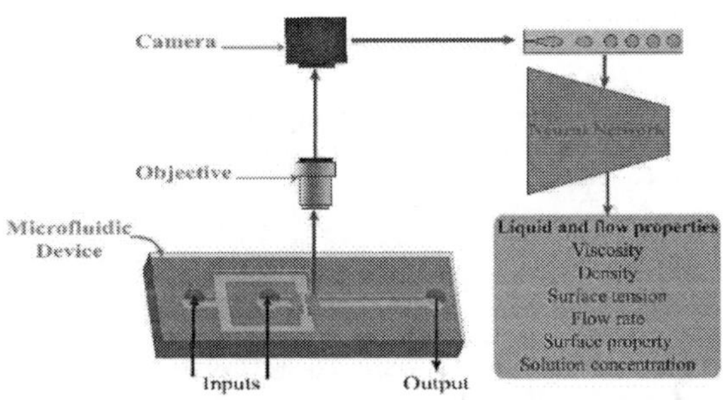

Fig. 1. Images of droplet formation in a microfluidic channel are used to train a neural network to extract information about the liquid and flow properties.

2 Methods

A flow focusing geometry is used to generate water-IPA droplets in silicone oil. The microfluidic device is fabricated in PDMS and bonded to glass. The flow rates are controlled using two syringe pumps. Images of droplet flows are captured using a microscope and a high-speed camera. For the flow rate measurement, the flow rate varies from 0.1ml/h to 1.5 ml/h with 0.05ml/h step. In this experiment the concentration of water-IPA is at constant concentration of 5.5%. For the concentration measurement, the concentration of IPA in water is changing from 4% to 7% with 0.5% pitch. A regression neural network shown in Fig. 2 is implemented in Tensorflow library for the flow rate and concentration measurement. The network has three sets of convolutional/max pooling layer following by a fully connected layer. This network has a single output which gives the fluid property.

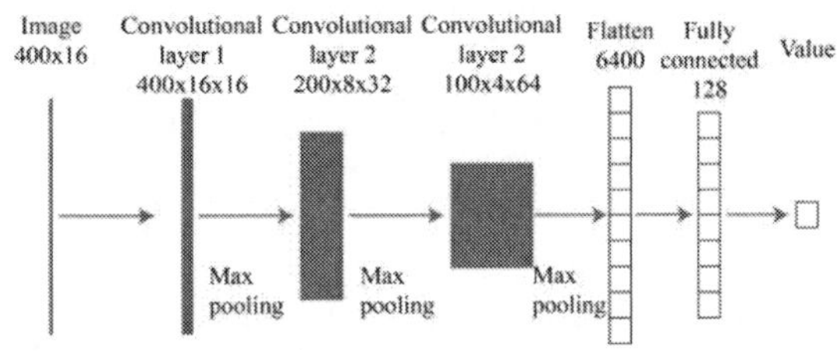

Fig. 2. Structure of neural network.

3 Results

Fig. 3a shows some instances of droplet flows at different flow rates. Images of flow rates from 0.1ml/h to 1.5ml/h with 0.1ml/h step are used to train the network. Afterwards, new images of the same range of flow rate with 0.05ml/h pitch are used for the test. Error of measured values by the DNN is shown in Fig. 3b. The mean error of the measurement for the trained and new flow rates are 2.878% and 5.729%, respectively. The results show the capability of the DNN in interpolating between the trained values for the cases that is new for the network. For new flow rates, there are some cases that the measured value is different from the true value which is caused by sudden change of droplet flow regime.

* Corresponding author: pooria.hadikhani@epfl.ch

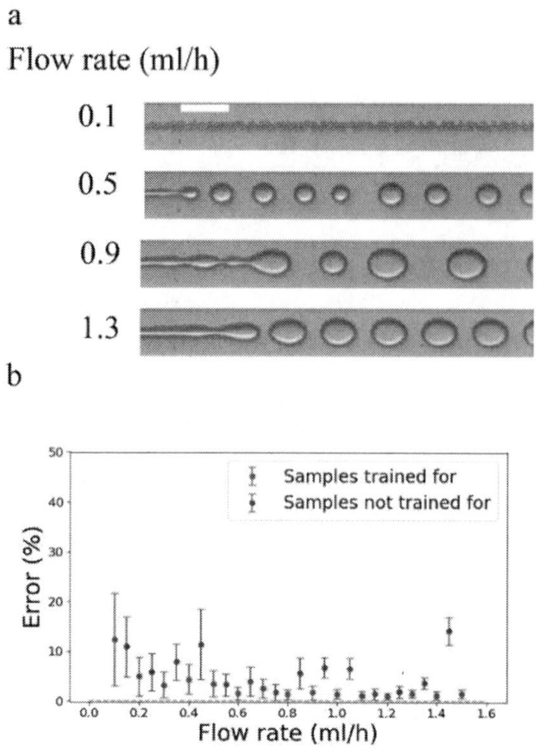

Fig. 3. Flow rate measurement: (a) Images of droplets at different flow rates. (b) Error of measured values by the network at different flow rates.

As can be seen in Fig. 4a, the concentration of isopropanol in the water is affecting the droplet flow pattern. The neural network is trained with samples ranging from 4.0% to 7.0% with 1.0% step. The network is tested with concentrations in the same range but with 0.5% steps. Due to a flow regime transition between concentrations of 5.0% and 6.0%, the mean error of measurement for the new concentrations is 9.306%. Because the flow regime transition is dependent on velocity as well as the concentration, we trained six networks at six different velocities to measure the concentration. Maximum three outliers of the measured values are removed and the closest measurement to the average is considered as the concentration. Furthermore, the output is calculated by averaging over the measurements for ten sample images input. This procedure decreases the mean error of measurement for the concentrations outside of the training set to 2.583%. Fig. 4b shows the error of measurement at each concentration.

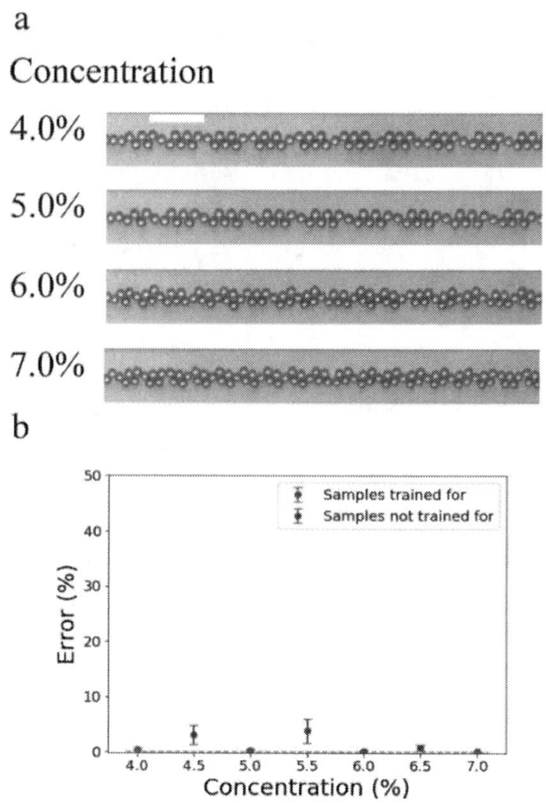

Fig. 4. Concentration measurement: (a) Images of droplets at different concentrations. (b) Error of measured values by the network at different concentrations.

4 Conclusion

In this study, we used deep neural networks to extract information about the flow/liquid properties from droplet flow images. As an example, DNNs are used to measure the flow rate of the flow and the concentration of the liquid components. DNNs can be used to discover other fluid/flow properties such as viscosity, surface tension, material surface property, and density. This method has potential to be extended to measure multiple fluid properties from a single image window on a microfluidic chip.

Microfluidic engineering for continuous in-flow cyto-tomography

Pasquale Memmolo[1,*], *Massimiliano M.* Villone[2], *Francesco* Merola[1], *Lisa* Miccio[1], *Martina* Mugnano[1], *Pier Luca* Maffettone[2], and *Pietro* Ferraro[1]

[1]National Research Council of Italy, Institute of Applied Sciences and Intelligent Systems "E. Caianiello", Via Campi Flegrei 34, Pozzuoli, Naples, Italy.
[2]Dipartimento di Ingegneria Chimica, dei Materiali e della Produzione Industriale, University of Naples "Federico II", Piazzale Tecchio 80, 80125 Napoli, Italy.

Abstract. The possibility to investigate cells in microfluidic flow by using a full 3D morphometry analysis is highly demanded to achieve information about their healthiness. Recently, the tomographic flow cytometry by digital holography has been demonstrated to monitor red blood cells in microfluidics environment by simply applying flux pressure to induce random self-rotation of flowing cells. Here, we provide a microfluidic solution to engineer the flow with the aim to ensure the full 360 degree of angle rotation of all cells in the field of view. We test the proposed methods for circulating tumour cells.

1 Introduction

Quantitative phase imaging by digital holography is one of the most powerful imaging solution to investigate living biological samples [1-3]. The quantitative phase-contrast images, that can be recovered by holographic imaging in microscope configuration, encode the combination of two biophysical information of imaged cells, i.e. their refractive index and physical thickness. Therefore, it is possible to measure them only after a suitable decoupling procedure [4,5]. On the other hand, tomographic phase microscopy (TPM) [6,7] permits to retrieve the 3D refractive index mapping of cells, when a multiple observation from different views is implemented. Actually, three main approaches are known to image cells from different directions, i.e. beam deflaction [8], mechanical rotation [9] and microfluidic induced rotation [10,11].
TPM by beam deflection suffers from some limitations, mainly related to the limited angle scanning range (up to 180 degrees), while TPM by mechanical rotation is intrinsically invasive, due to the need of manipulating cells. Actually, a different solution that uses an external force to manipulate cells have been proposed in ref. [12], where optical tweezers are used to trap and rotate cells. Of course, this method needs to a more complicate optical arrangement related to the use of a second light source to implemented the optical trapping [13].
Recently, the Tomographic flow cytometry (TFC) has been demonstrated [10], which overcome the limitation of the angle scanning range, providing the full 360 degrees of cells rotation, along with standard optical set-up. In fact, it uses digital holography in microscopy as the imaging tool and a microfluidic channel in which cells move in flow. TFC is successfully applied to reconstruct the tomogram of red blood cells and diatoms. However, the induced rotation of cell is not controlled, thus limiting the throughput of the proposed tomography system. Here we show a different solution to overcome this problem, by a microfluidic engineering of the flow through the fluid dynamic modelling. In particular, we design the physical sizes of the microfluidic channel along with the optimal condition of the pressure pump to ensure the cells tumbling, i.e. the rotation of cells without mechanical deformation [11]. In this case, the proposed TFC is suitable for the continuous in-flow cyto-tomography.

2 Results and discussion

The TFC by digital holography apparatus consist of a holographic microscope which makes imaging of flowing cells in a microfluidic channel. The cells tumbling is induced by the external force, through a controlled pressure pump. The cells moves and rotate continuously in the imaged field of view and the holographic recording with high frame rate is performed. After recording of video sequence of digital holograms, the numerical processing is implemented to recover the quantitative phase-contrast images. In particular, the automatic focusing algorithm is used to retrieve the axial position of each cells in the field of view, then the holographic 3D tracking [14] is employed to realign the detected cells and a suitable algorithm is used to calculate the tumbling angles frame-by-frame [10,11]. Finally, the tomogram is obtained by using the optical projection tomography method. In figure 1, the complete pipeline of the TFC is reported with an example of application in the case of circulating breast cancer cells.

* Corresponding author: p.memmolo@isasi.cnr.it

 https://doi.org/10.1051/epjconf/201921510003

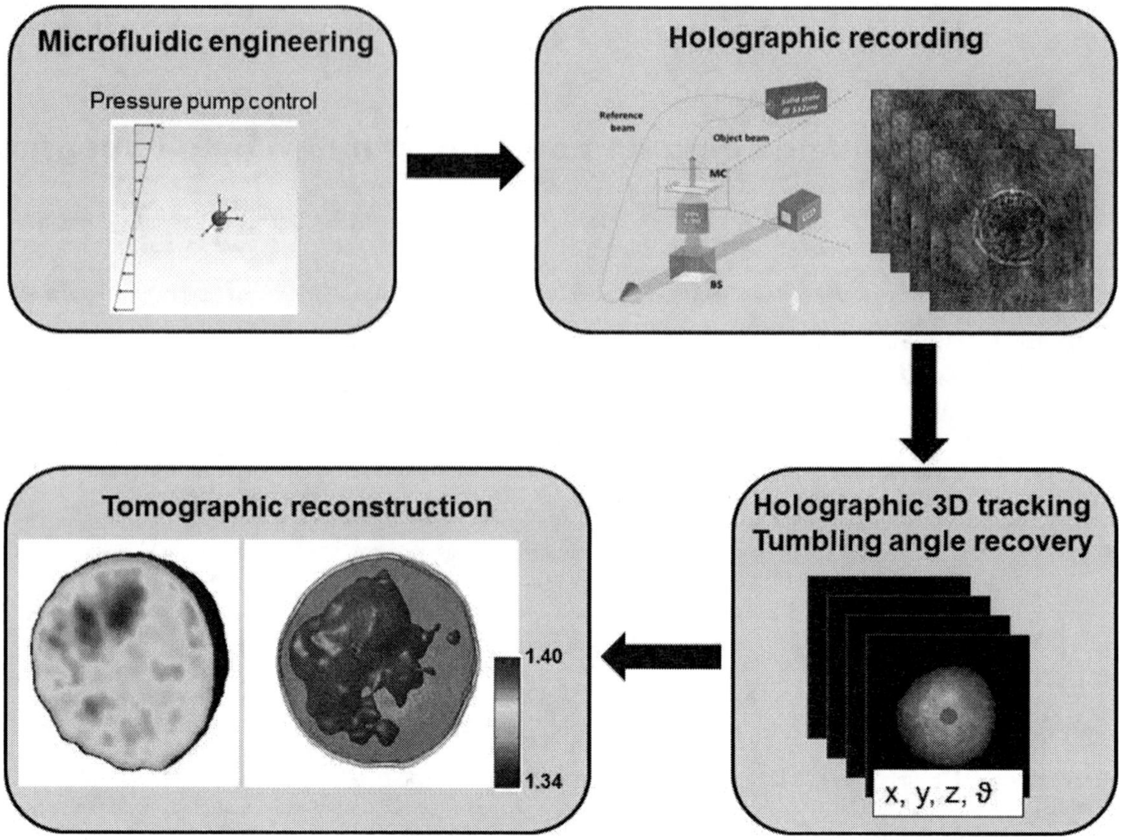

Fig. 1. Continuous TFC by microfluidic engineering.

References

1. P. Ferraro, A. Wax, Z. Zalevsky, *Coherent Light Microscopy* (Series in Surface Sciences, Springer 46 2011).
2. Y.K. Park, C. Depeursinge, G. Popescu, Nat Photon **12**, 578 - 589 (2018).
3. F. Merola, P. Memmolo, L. Miccio, V. Bianco, M. Paturzo, P. Ferraro, Proc. IEEE **103**(2), 192 - 204 (2015).
4. B. Kemper, I. Bredebusch, W. Domschke, S. Kosmeier, P. Langehanenberg, J. Schnekenburger, G. von Bally, J. Biomed Opt **12**, 054009 (2007).
5. P. Memmolo, L. Miccio, F. Merola, O. Gennari, P.A. Netti, P. Ferraro, Cytometry Part A **85A**, 1030-1036 (2014).
6. D. Jin, R. Zhou, Z. Yaqoob, P.T.C. So, J. Opt. Soc. Am. B **34**, B64-B77 (2017).
7. F. Merola, P. Memmolo, L. Miccio, M. Mugnano, P. Ferraro, Methods **136**, 108-115 (2018).
8. W. Choi, C. Fang-Yen, K. Badizadegan, S. Oh, N. Lue, R.R. Dasari, M.S. Feld, Nat Methods **4**, 717–719 (2007).
9. F. Charrière, A. Marian, F. Montfort, J. Kuehn, T. Colomb, E. Cuche, P. Marquet, C. Depeursinge, Opt Lett **31**, 178-180 (2006).
10. F. Merola, P. Memmolo, L. Miccio, R. Savoia, M. Mugnano, A. Fontana, G. D'Ippolito, A. Sardo, A. Iolascon, A. Gambale, P. Ferraro, Light Sci. Appl. **6**, e16241 (2017).
11. M.M. Villone, P. Memmolo, F. Merola, M. Mugnano, L. Miccio, P.L. Maffettone, P. Ferraro, Lab Chip **18**, 126-131 (2018).
12. M. Habaza, B. Gilboa, Y. Roichman, N.T. Shaked, Opt Lett **40**, 1881-1884 (2015).
13. M. Padgett, R. Di Leonardo, Lab Chip **11**, 1196–1205 (2011).
14. P. Memmolo, L. Miccio, M. Paturzo, G. Di Caprio, G. Coppola, P.A. Netti, P. Ferraro, Adv Opt Photon **7**, 713-755 (2015).

 https://doi.org/10.1051/epjconf/201921511001

Guided Mode Resonances For Sensing And Imaging

Jose Juan-Colas, *Donato* Conteduca, *Isabel* Barth *and Thomas F* Krauss*

Department of Physics, University of York, York, YO10 5DD

Abstract. We discuss the latest developments in guided mode resonances for sensing and imaging and compare them to alternative approaches such as plasmonic nanohole arrays.

1 Introduction

Resonant nanophotonic concepts offer interesting opportunities for sensing and imaging applications. These concepts include microring resonators, gratings/photonic crystals as well as plasmonic nanostructures such as nanoparticles and nanohole arrays. Our work focuses on guided mode resonances (GMRs) in dielectric arrays. Such guided mode resonances in the context of sensing were first described by Magnusson [1] and then further developed by Cunningham, who referred to them as photonic crystal biosensors [2].

2 Sensing

The key advantage of the guided mode resonance approach is that GMRs are easily excited by out-of plane illumination using simple collimated light sources, including LEDs. This ease of coupling allows them to be used in low-cost packages, including integration into removable cartridges that are inserted by untrained personnel for point-of-care applications. An artist's impression of a possible realisation is shown in fig. 1. The ease of coupling is due to the fact that GMRs are leaky modes, which means that they do not achieve the high Q-factors achievable with guided-mode optics devices, but that they can be readily excited with a collimated beam with large alignment tolerances. The challenge, therefore, is to achieve high performance while retaining the advantage of simple interfacing. An additional challenge is the readout, which in many cases requires an external spectrometer.

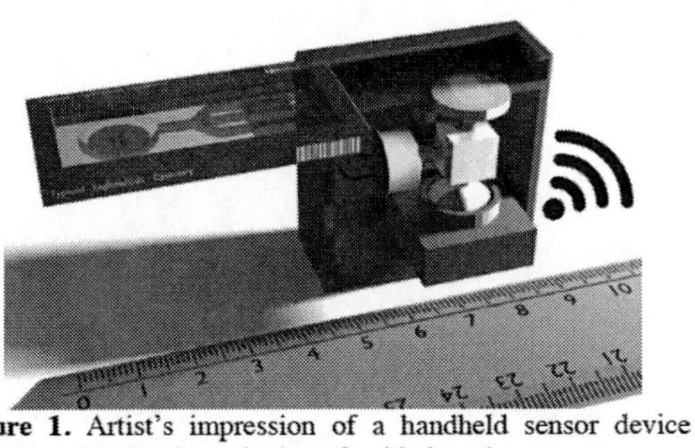

Figure 1. Artist's impression of a handheld sensor device exploiting the simple excitation of guided mode resonances.

By chirping the period or fill-factor of the GMR [3], we have been able to solve the spectrometer problem by integrating the sensing and the readout function into the same structure.

Inspired by the success of sensing and imaging with plasmonic nanohole arrays [4], we have also explored nanohole arrays in silicon (fig. 2a) and have obtained remarkable results; in particular, we have observed beautiful Fano resonances with a Q-factor of 200-300 (fig. 2b, factor 5-10 higher than a plasmonic resonance) as well as a very high surface sensitivity which is comparable to that of the plasmonic equivalent.

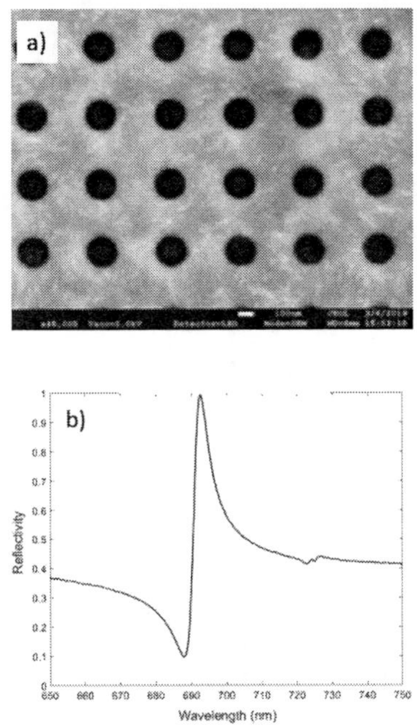

Figure 2. a) Nanohole array of 470 nm period realised in hydrogenated amorphous silicon. b) Corresponding Fano resonance. Note the large dynamic range and the near-100% reflectivity at a wavelength below 700 nm, which highlights the low loss of the hydrogenated a:Si material.

Finally, we are now implementing common-path interferometric methods which further increase the available sensitivity. As a result, our detection limit is now in the $\Delta n \approx 1e\text{-}6$ regime, which is comparable to much

* Corresponding author: thomas.krauss@york.ac.uk

more sophisticated approaches yet still compatible with a handheld device.

3 Imaging

The idea of the chirped GMR approach is that it converts spectral information into spatial information that can be easily picked up by a simple camera, i.e. it exploits the imaging capability of the GMR; one can therefore also think of a GMR as an array of sensors. Given the high sensitivity, it should then be possible to detect proteins locally, i.e. in the proximity of a cell. Taking this idea further, we have used the technique to image human cells and the proteins they secrete, which allows us to see how cells communicate in health and disease. Since the technique is label-free, cells can be directly taken from patients with minimum preparation. We used the technique to map the secretion of the signaling protein thrombopoietin (TPO) from individual human HepG2 cells and quantified the heterogeneity of TPO expression as a function of the desialylated platelet concentration (fig. 3) [5].

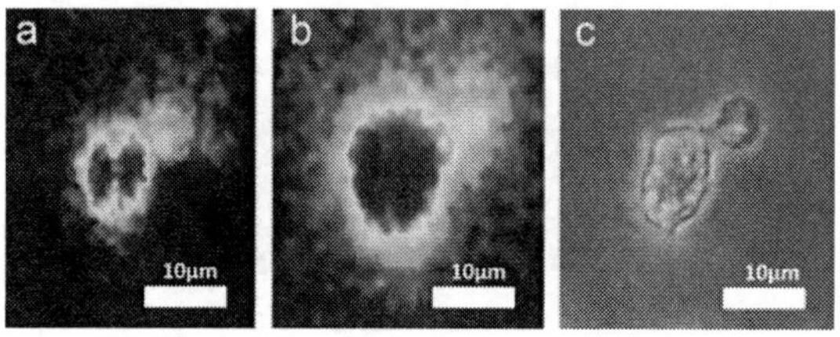

Figure 3. a) Human HePpG2 cell imaged at the beginning of the experiment (t=0 min). b) The same cell at t=84 min. It appears larger because it has secreted thrombopoietin (TPO). The same cell imaged by phase contrast microscopy; the image does not change during the same experiment. The spatial resolution is higher but phase contrast is not able to see protein secretion.

4 Conclusion

Overall, the versatility of the GMR approach makes it very attractive for both sensing and imaging applications. It also offers multiple degrees of freedom which offer opportunities for further optimization.

References

[1] R. Magnusson, S.S. Wang, APL **61** pp. 1022-1024 (1992).

[2] B. Cunningham, P. Li, B. Lin, J. Pepper, Sens & Act **B 81**, pp. 316-328 (2002).

[3] G.J. Triggs, Y. Wang, C.P. Reardon, M. Fischer, G.J.O. Evans, T.F. Krauss, Optica **4**, pp. 229-234 (2017).

[4] X. Li, M. Soler, C.I. Özdemir, A. Belushkin, F. Yesilköy H. Altug, Lab Chip **17**, pp. 2208–2217 (2017).

[5] J. Juan-Colás, I.S. Hitchcock, M. Coles, S. Johnson, T.F. Krauss, PNAS **115** pp. 13204–13209 (2018).

Optofluidic lab-on-chips for high throughput 3D imaging of cells and tissues

Federico Sala,[1;2], *Petra* Paiè,[1] , *Roberto* Memeo,[2], *Roberto* Osellame,[1;2], *Andrea* Farina,[1], *Andrea* Bassi,[2] *and Francesca* Bragheri[1]

[1]Istituto di Fotonica e Nanotecnologie - Consiglio Nazionale delle Ricerche (IFN-CNR), 20133 Milano, Italy
[2]Dipartimento di Fisica - Politecnico di Milano, 20133 Milano, Italy

Abstract. Optofludic lab-on-chips are compact microsystems that allow the manipulation, the analysis and the imaging of cells with high throughput even at the single cell level. The optofluidic microscopes on chip here presented are fabricated by femtosecond laser micromachining followed by chemical etching. The technique is ideally suited for rapid prototyping and allows 3D geometries and perfect alignment between microfluidic and optical components networks. The devices have been validated on cancer cell lines by imaging in three dimensions both clusters of cells growing on spheres and single cells.

1 Introduction

Recent advances in imaging technologies, as the enhancement of spatial and temporal resolution, push the limits of the research in biology by extending the imaging capabilities from the surface of cell samples deeper into organisms. However, one major challenge is to provide accurate statistical analysis that takes into account the heterogeneity of cellular behaviour in the same biological model.

Optofluidics provides a perfect synergy for the development of new imaging platforms able to perform recording and analysis of multiple samples with minimal photodamage, by the combination of microfluidics and optics. Optofluidic devices or lab-on-chip can thus be used for cell analysis and imaging either at the single cell level to study heterogeneity of specific cell properties in uniform populations or when cells during proliferation form a spheroid or even an organoid.

Three dimensional optical microscopy techniques, such as light sheet microscopy [1], have shown to be a powerful tool to rapidly image biological samples at high spatial resolution, nevertheless complex sample preparation and system alignment still limits the imaging of multiple specimen. Few integrated approaches have been proposed [2], and we recently demonstrated a selective plane microscope on-chip [3] that permits to analyse cellular spheroids at high throughput.

Here we report the fabrication and the experimental validation of a class of optofluidic light sheet fluorescent microscopes on-chip that exploit microfluidics for easy sample movements and laser light for the excitation of the samples under investigation. We show a new class of microscopes on chip where the optical and fluidic components have been designed to allow for dual-colour high throughput 3D imaging of cellular spheroids and single cells. The devices have been fabricated by femtosecond laser micromachining followed by chemical etching, which is ideally suited for rapid prototyping of optofluidic glass chip thanks to its 3D capability in irradiating both optical components as waveguides or lenses and microchannels. Laser writing is a serial fabrication technique that does not allow parallel fabrication, thus limiting the production yield. On the other hand it permits easily to modify the chip design in order to customize and optimize the devices. This feature has been exploited to realize microscope on chip for the 3D imaging of both tumour spheroids and single cancer cells.

2 Device fabrication

Femtosecond laser micromachining followed by chemical etching (FLICE) is exploited to fabricate both the optical and the fluidic components of the lab-on-chips (LOCs). The technique is a two-step fabrication process consisting of: 1) permanent material modification due to nonlinear absorption of focused femtosecond laser pulses; 2) etching of the laser modified zone by a hydrofluoric acid (HF) solution. The laser irradiation enhances the etching rate in comparison to the pristine material, thus channels with arbitrary shape can be obtained in bulk glass substrates.

The integrated selective plane illumination microscope is composed by two embedded cylindrical lenses that are used to focus the laser light for imaging. Two light sheets that cross a C-shaped microchannel from two opposite sides are thus formed. This

* Corresponding author: francesca.bragheri@ifn.cnr.it

microchannel has a square cross-section and allows the sample scanning by letting the samples flow at a constant speed through the light sheets. The cylindrical lenses are realized with a microfluidic channel having an engineered cross-section to optimize aberrations and filled with a high refractive index fluid. A schematic of the device is shown in Fig.1 (left). The complicated microchannel network necessary to obtain both the lenses and the microchannel for sample delivery can be easily realized in a single irradiation step by exploiting the 3D capability of the FLICE technique.

Thanks to the versatility of FLICE technique, we have also fabricated prototypes where cylindrical lenses are empty and realized with the air/glass interface having an engineered profile to avoid aberrations of the focused light. The microchannel shape has also been modified, as the imaging of single cell requires a higher numerical aperture objective for the detection. A further optimization of the device has been introduced by inscribing in the substrate two waveguides facing the two lenses, which allows a precise alignment of the two light sheets in the microchannel. Filters have been also fabricated by FLICE in order to remove the light that is not coupled to the waveguide and to increase the signal to noise ratio of the illumination planes.

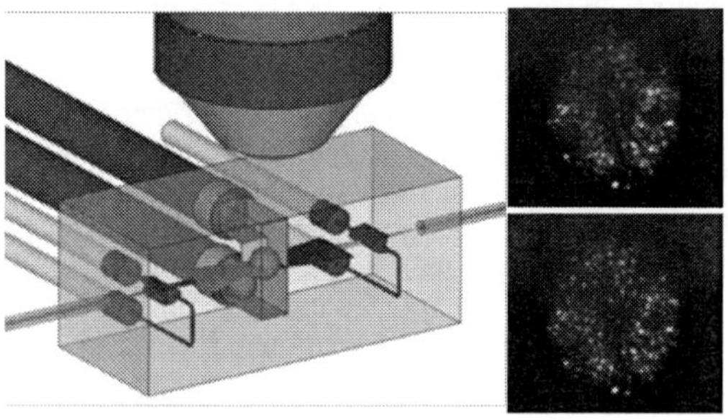

Fig. 1. Schematic of a dual-colour optofluidic selective plane illumination microscope on chip. An example of maximum intensity projection image of a tumour spheroid is shown on the right along with a segmentation of the image where the nuclei of cells are highlighted with different colours.

3 Device characterization

The engineering of the devices allowed obtaining different thicknesses of the light sheets ranging from 1.5 μm to 7 μm, which can be alternatively implemented on the chip on the basis of the sample to be analysed. For the microscope on chip validation fixed tissue spheroids of colon cancer cells expressing a histone H2B fluorescent nuclear reporter protein (mCherry) has been used. They were flown in the microchannel at a constant speed passing through the light sheet that excited fluorescence of a spheroid section at time. By collecting the emitted fluorescence the entire 3D image could then be reconstructed. An example of the Maximum Intensity Projection of a spheroid and its segmentation analysis are shown in Fig.1(right). The microfluidic microscope for single cell analysis has been validated with fixed breast cancer cells, demonstrating the capability of this device to obtain 3D reconstruction of the sample with very high resolution. High throughput acquisition is also demonstrated paving the way for a statistically significant method of single cell analysis by high-resolution 3D imaging.

The results showed the possibility to obtain important biological insights in the modification of cellular properties during tumour progression. The increase in the achievable resolution would allow for imaging of the chromatin network so as to evidence modifications in the structure that might be associated to cancer features. All these advancement would open new possibilities towards the field of personalized medicine

This work was partially supported by the European Union under the FET Open PROCHIP project (grant agreement no. 801336).

References

1. J. Huisken, J. Swoger, F. Del Bene, J. Wittbrodt, E.H.K. Stelzer, Science **305**, 1007 (2004)
2. P. Paiè, R. Martinez Vzquez, R. Osellame, F. Bragheri, A. Bassi, Cytometry Part A **93**, 987 (2018)
3. P. Paiè, F. Bragheri, A. Bassi and R. Osellame, Lab on a Chip **16**, 1556 (2016)
4. R. Osellame, H. J. W. M. Hoekstra, G. Cerullo, M. Pollnau, Laser & Photonics Reviews **5**, 442, (2011)

Advanced Optical Imaging of Blood Thrombus

Xuefei He[1] , *Samantha J. Montague*[1,2] , *Xu Tao*[1], *Elizabeth E. Gardiner*[1], *Woei Ming Lee* [1,2,3,*]

[1] Research School of Electrical, Energy and Materials Engineering,, College of Engineering and Computer Science, The Australian National University, Canberra ACT 2601, Australia
[2] ACRF Department of Cancer Biology and Therapeutics, John Curtin School of Medical Research, The Australian National University, Canberra ACT 2601, Australia
[3] The ARC Centre of Excellence in Advanced Molecular Imaging, The Australian National University, ACT 2601 Australia

Abstract. The application of advanced optical imaging technology for platelet biology is in its infancy. In this talk, we shall introduce the potential impact of optical imaging in understanding the roles of platelets in thrombus formation and in hemostasis. We will summarize how techniques in optical imaging has enabled exploration of morphological, molecular and fluidic parameters that determine platelet adhesion, activation and consequent roles in thrombus formation. Finally, we discuss the convergence of multiple imaging modalities towards a complete understanding of platelet roles in thrombus formation.

1. Thrombus formatin

Discoveries in biological sciences have largely been driven by the advances in technological tools. The first observation of platelets were made by Max Schultze who used a 17th century optical microscope with a moist chamber to reveal individual granulates (particles) during the process of coagulation (formation of a blood clot). However, this method was insufficient for in depth investigations. It was only until the use of immersion microscope objective by Bizzozero, an Italian microscopist, in 1881 that lead to an in-depth characterization of platelets as an independent component of blood cells around the mesenteric vessels of rabbits and guinea pigs. A key finding by Bizzozero was the crucial role of platelets in the formation of hemostatic thrombus following a controlled injury to blood vessels.

Since then, platelets have been established as a crucial component of blood due to their fundamental roles in hemostasis (process of stopping blood flow). Studies of platelets are generally focused on deeper understanding of the finely regulated molecular mechanism in hemostasis and thrombus formation (process of blood clot), which have major involvement in o different diseases progressions such as atherosclerosis, angiogenesis and tumorigenesis. Recently, platelets have also been closely associated with inflammation, responses to infection in the immune system and vascular integrity.

Platelets are normally in a quiescent state (inactive) in the blood circulation, until it is delicately triggered to response to sudden fluid shear stress and/or exposure to extracellular matrix (ECM) proteins on the walls of damaged vessels It is the immediate mechanical responsiveness of platelets towards activation and adhesion to sites of injury anjd the smallest changes or disturbances to regular laminar flow that makes it profoundly difficult to quantify in vivo [1, 2]. There have been numerous in vitro systems designed to measure the adhesive and activating events of platelets during thrombus formation. However, , the innate mechanisms that control thrombus propagation, thrombus stability and embolism in an in vivo setting remains poorly quantified and understood.[3, 4].

2. Imaging thrombus under flow using microfluidics

The utilization of microfluidic devices for biological studies has developed into a popular field that allows clear ex vivo observation and measurement of thrombus formation, but still maintains the basic physiological environment by surface modification, shear stress control or different geometrical designs. In recent years, the development of microfluidic techniques has merged into a versatile tool for investigating platelet functions and roles of specific platelet receptor and response to different types of agonists, including collagen or shear conditions (references). Along with this, various kinds of flow assays have been proposed for fast and low cost diagnosis in clinical settings. Although the microfluidic device offers a highly flexible and stable platform for platelet and thrombosis studies, there are still insufficiencies and limitations. A challenge in this field is that the lack of standardization. Different microfluidic channels designs, materials used and devices that results in inconsistency. Furthermore, imaging instruments that can accurately quantify physical and molecular parameters of platelets during the thrombosis processes as well as the changes in the microenvironment with in the body would be a game changer in platelet biology. To better dissect the processes that imaging can quantify, we

* Corresponding author: steve.lee@anu.edu.au

separate the challenges into three broad imaging categories namely: morphological, molecular, and fluidic. Morphological information refers to accurate mapping of the physical dimension of a developing thrombus. Molecular imaging refers to individual platelets surface protein expression and signalling events involved at the different stages of activation. Finally, real time recording of fluidic flow patterns or shear rate manipulation over the developing thrombus would provide the ability to localise the effects of a shear gradient in situ.

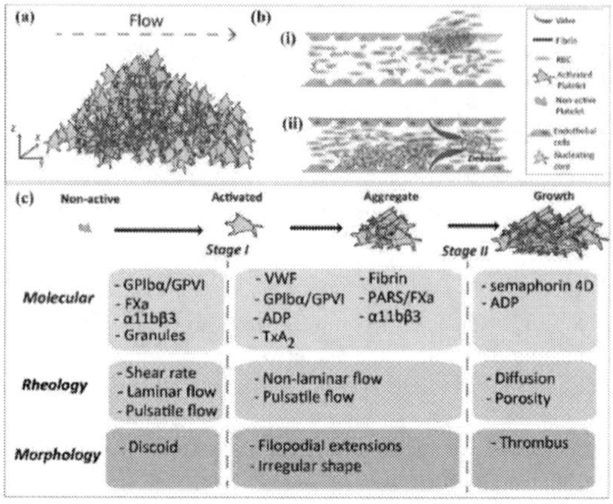

Figure 1. Shows the different molecular, rheology and morphological events that needs to be imaged under flow to study blood thrombus.

3. Multimodality imaging in thrombus and platelet studies

There are over half a million platelets in a single microliter of blood and each platelet are intricately crucial to the role of haemostasis; the ability to arrest blood loss after traumatic injury. There is crucial need to develop a fundamental understanding of how platelets respond to variations in changing shear gradient at a molecular and morphology level. Hence the simultaneous quantification of platelet thrombus formation and platelet receptor expression and shedding with precise determination of shear rates to micrometer levels across an individual thrombus, (on the same sample) will enable shear-dependent effects on molecular events to be directly correlated. There are therefore enormous clinical and therapeutic implications for these shear-related events, and for the role of altered vascular rheology and shear in thrombus formation in diseased vessels. All in all, a comprehensive imaging system that can directly records the morphology, molecular and fluidic of a developing thrombus under flow will have a paradigm shift in platelet biology. An example of an effective use of multi-scale imaging for platelet studies was conducted by Gaertner and colleagues [5]. They have shown that platelet have an innate ability to detect invasion of foreign entities along the vascular bed using a series of imaging tools (fluorescence, FIB-SEM, multiphoton microscopy and differential interference contrast) to correlate the morphological and molecular from microscale to nanoscale. By gathering evidence through these imaging tools, they demonstrated that platelets have the strength to overcome adhesion forces and migrate towards invading bacteria. This extraordinary finding rests on the hypothesis that membrane adhesion receptors of adhering platelets can actively probe the mechanical properties of an adhesive substrate. Hence, the ability to combine combining two or more complementary imaging tools for platelets and thrombus studies could produce multiple viewpoints of a developing thrombus by simultaneously capturing biological activities at single platelet level and tracking the overall packing density of thrombus under flow. For example, it is easily conceivable for one to combine volumetric imaging techniques: quantitative phase microscopy and lightsheet to investigate the morphology of the thrombus and platelets volumetrically. This is especially because light sheet microscopy uses very low density of light and causes almost no adverse phototoxic outcomes. The lightsheet component of the microscopy system will be able to capture individual calcium signals of platelets interacting and at the same time a QPM [6] can directly quantifies the overall growth of the thrombus under continuous fluid flows. The simultaneous acquisition of these parameters will allow investigators to study the interplay between micro-rheological and agonist-dependent mechanisms of heterogeneous platelet activation within a developing thrombus with un-parallel clarity and resolution not previously obtained. While current optical imaging studies using multiphoton microscope in vivo [7] have shown evidence of heterogeneous molecular signatures between platelets within a single thrombus i.e. fully activated, the ability to quantify individual platelet interactions in a large thrombus at a sub-cellular level under flow is a grand challenge. In conclusion, the increasing use of advanced imaging techniques in platelet and thrombus studies would inevitable increase our understanding of thrombotic processes that occurs at short time intervals (milliseconds to seconds) on a micrometre size scale and with rapidly changing fluidic flows. We anticipate that in situ fluidic measurements combined with optical imaging techniques would be a necessary addition for the next generation of platelet imaging tools

References

1. Dreymueller, D., S. Uhlig, and A. Ludwig, Am J Physiol Lung Cell Mol Physiol, 2015. **308**(4): p. L325-43.
2. Coenen, D.M., T.G. Mastenbroek, and J.M.E.M. Cosemans, Blood, 2017. **130**(26): p. 2819-2828.
3. Tomaiuolo, M., L.F. Brass, and T.J. Stalker, Interv Cardiol Clin, 2017. **6**(1): p. 1-12.
4. Jackson, S.P., Blood, 2007. **109**(12): p. 5087-5095.
5. Gaertner, F., et al., Cell, 2017. **171**(6): p. 1368-1382.e23.
6. He, X., et al., Advanced Biosystems, 2018. **2**(10): p. 1800089.
7. Li, Y., et al., Journal of Biophotonics, 2018. **11**(7): p. e201700341.

Nanophotonic Metasurfaces for Biosensing and Imaging

Hatice Altug[1]

[1]EPFL, Ecole Polytechnique Federale de Lausanne, School of Engineering, Lausanne, Switzerland, 1015

https://bios.epfl.ch/

Abstract. Nanophotonics excels at confining light into nanoscale optical mode volumes and generating dramatically enhanced light matter interactions. These unique aspects have been unveiling a plethora of fundamentally new optical phenomena, yet a critical issue ahead for nanophotonics is the development of novel devices and applications that can take advantage of these nano-scale effects. It is expected that nanophotonics will lead to disruptive technologies in energy harvesting, quantum and integrated photonics, optical computing and including biosensing. To this end, our research is focused on the application of nanophotonics to introduce powerful biosensors that can have impact on a wide range of areas including basic research in life sciences, early disease diagnostics, safety and point-of-care testing. In particular, we exploit nanophotonics and its integration with microfluidics to address key challenges of current biosensors and develop devices that can enable label-free, ultra-sensitive, multiplexed, rapid and real-time measurements on biomolecules, pathogens and living systems. In this talk I will present some of our recent work on nanophotonic metasurfaces for biosensing and bioimaging as well as their applications in real-world settings.

1 Introduction

With the emerging new healthcare initiatives including personalized medicine, global health care and point-of-care diagnostics, we are witnessing a paradigm shift in the way medical care is provided. These initiatives are demanding next generation diagnostics and bioanalytical tools that can deliver a long list of desired specifications such as high sensitivity, multiplexing, real-time operation, and miniaturization, low-cost. Nanophotonics offers unique opportunities to meet such requirements. To this end the effort of our lab is focused on using nanophotonics to develop powerful biosensors can impact numerous fields ranging from basic research in life sciences, early disease diagnostics, personalized medicine and point-of-care testing to environmental/food monitoring and safety. In particular we employ a variety of nanophotonic techniques including dielectric metasurfaces and plasmonics to introduce fundamentally new optical detection schemes. We exploit electromagnetic spectrum spanning from visible to mid-infrared wavelengths for developing biosensors based on optical refractive index variations and surface enhanced IR absorption spectroscopy. We integrate our sensors with microfluidic systems for efficient analyte trapping, sample manipulation and automation. We also combine the use of smart data science tools with bioimaging based sensing architectures for achieving unprecedented performance metrics. In this talk, I will present some of our recent work on nanophotonic metasurfaces for biosensing and bioimaging and show examples how they can be used in real-world applications [1-14].

2 Nanophotonic Biosensors and Imaging Systems

2.1 Mid-IR Plasmonics for Label-Free, Sensitive and Real-Time Monitoring of Biomolecular Conformational Changes

We use engineered Mid-IR plasmonic antennas to overcome fundamental limitations of conventional IR spectroscopy by surface enhanced infrared absorption (SEIRA) spectroscopy method. Recently, we showed a plasmonic metasurface consisting of multi-resonant nanoantennas for simultaneously distinguishing multiple analytes in heterogeneous biological samples [2]. We demonstrated that this device can spectroscopically resolve the interaction of biomimetic lipid membranes with peptides as well as the dynamics of synaptic vesical cargo release. Our lab has been also advancing label-free spectroscopic identification of secondary structure conformation of nanometric thin protein layers in aqueous solution. Detection of protein conformations is immensely important to understand and diagnose incurable diseases resulting from protein misfolding such as Parkinson's and Alzheimer's disease. In our 2017 work, we used α-synuclein, a protein associated with plaque formation in Parkinson's diseases [3]. By leveraging plasmonically enhanced amide-band proteins, we were able to measure a random coil to cross β-sheet conformational change as a result of α-synuclein aggregation. We showed high structural sensitivity of our

* Corresponding author: hatice.altug@epfl.ch

biosensor by distinguishing the characteristic β-sheet conformations between a native protein and pathological fibrillar aggregates. In a recent work, we presented the first experimental demonstration of real-time label-free spectroscopic monitoring of secondary structure changes in a protein monolayer in aqueous solution [4]. We have been contributing to the fundamental understanding of SEIRA and determining its limitations. Since SEIRA is a new field, the general notion on sensitivity parameters is still not well developed. In a recent work, we introduced methodologies in order to quantify the limit of detection which can provide a valuable reference points for determining biosensor performance metrics [5].

2.2 All-Dielectric Metasurfaces for Biosensensing and Imaging

All-dielectric metasurfaces are fast emerging in nanophotonics field due to their low intrinsic loss, strong resonance enhancement of both electric and magnetic near-fields and complementary metal-oxide semiconductor (CMOS) compatibility. In 2018 [6], we introduced a nanophotonic method capable of detecting mid-infrared molecular fingerprints without the need for spectrometry, frequency scanning, or moving mechanical parts. Our method leverages a 2D array of high-Q dielectric metasurface pixels designed to provide strong near-field enhancement, where the resonance positions of individual metapixels are varied over a target mid-IR fingerprint range. Significantly, we assign each resonance position to a specific pixel of the metasurface, establishing a one-to-one mapping between spectral and spatial information. This unique configuration enables to sample near-field enhanced molecular absorption signatures of surface adsorbed molecules at multiple spectral points and to translate this information into a 2D barcode-like spatial absorption map for imaging-based detection. Recently we developed metasurfaces operating in visible spectrum, and by utilizing hyperspectral imaging and smart data science tools we are currently investigate how to unleash the full potentials of high-Q dielectric metasurfaces for achieving biosensors with unprecedented performance metrics [Yesilkoy et al. accepted 2019].

2.3 Opto-Fluidic Biosensors for Real-Time Secretion Analysis from Live Cells and Single Cells

Our lab is also working on the integration of nanophotonics and microfluidics to develop biosensors that can analyze live cells at the single cell resolution. In 2017, we introduced a nanophotonic biosensor enabling real-time monitoring of live cell secretions in a label-free manner [7]. Our microfluidic-integrated biosensor, compatible with common inverted microscopes can able to distinguish cell secretion events during long-term live cell culture. In 2018, we demonstrated a new technology for label-free monitoring of cell responses at the single cell sensitivity [8]. The device enables long term monitoring of isolated individual cells and identifying their secretion activity in real time, all without disturbing the cell environment or introducing invasive agents.

References

1. Oh and Altug "Performance Metrics and Enabling Technologies for Nanoplasmonic Biosensors", Nature Communications Vol. 9, p. 5263 (2018).
2. Rodrigo et al., "Resolving Molecule-Specific Information in Dynamic Lipid Membrane Processes with Multi-Resonant Infrared Metasurfaces", Nature Communications Vol. 9, p. 2160 (2018).
3. Etezadi et al., "Nanoplasmonic Mid-Infrared Biosensor for in-vitro Protein Secondary Structure Detection" Light Science and Application Vol.6 e17029 (2017).
4. Etezadi et al., "Real-Time in-Situ Secondary Structure Analysis of Protein Monolayer with Mid-Infrared Plasmonic Nanoantennas", ACS Sensors Vol. 3, p. 1109-1117 (2018).
5. John-Herpin et al. "Quantifying Limits of Detection of Surface-Enhanced Infrared Spectroscopy with Grating Order-Coupled Nanogap Antennas", ACS Photonics Vol. 5, p.4117-4124 (2018).
6. Tittl et al., "Imaging-Based Molecular Barcoding with Pixelated Dielectric Metasurfaces", Science Vol. 360, p. 1105-1109 (2018).
7. Li et al., "Plasmonic Nanohole Array Biosensor for Label-Free and Real-Time Analysis of Live Cell Secretion", Lab on a Chip Vol. 17, p. 2208-2217 (2017).
8. Li et al., "Label-Free Optofluidic Nanobiosensor Enables Real-Time Analysis of Single-Cell Cytokine Secretion", Small Vol. 14, 1870119 (2018).
9. Beluskin et al., "Nanoparticle-Enhanced Plasmonic Biosensor for Digital Biomarker Detection in a Microarray", ACS Nano Vol. 12, p. 4453-4461 (2018).
10. Yesilkoy et al. "Phase-Sensitive Plasmonic Biosensor Using a Portable and Large Field of View Interferometric Microarray Imager", Light Science and Application Vol. 7, p. 17152 (2018).
11. Soler et. al "Two-Dimensional Label-Free Affinity Analysis of Tumor Specific CD8 T Cells with a Biomimetic Plasmonic Sensor", ACS Sensors Vol. 3, p. 2286-2295 (2018).
12. Mohammadi et al. "Nanophotonic Platforms for Enhanced Chiral Sensing", ACS Photonics Vol, p. 2669-2675 (2018).
13. Gupta et al. "Self-assembly of nanostructured glass metasurfaces via templated fluid instabilities" Nature Nanotechnology (2019).
14. Rodrigo et al., "Mid-Infrared Plasmonic Biosensing with Graphene" Science Vol 349, pp. 165-168 (2015).

Fast and Accurate Thickness Mapping of Thin Liquid Films

Zhe Wang[1,2*], *Vincenzo Ferraro*[3], *Biagio Mandracchia*[1], *Ernesto Di Maio*[3], *Pier Luca Maffettone*[3] *and Pietro Ferraro*[1]

[1]CNR-ISASI, Istituto di Scienze Applicate e Sistemi Intelligenti «E. Caianiello» del CNR, Via Campi Flegrei 34, 80078 Pozzuoli, Napoli, Italy.
[2]College of Applied Sciences, Beijing University of Technology, 100124 Beijing, China.
[3]Dipartimento di Ingegneria Chimica, dei Materiali e della Produzione Industriale, Università di Napoli Federico II, Piazzale Tecchio 80, 80125 Napoli, Italy.

Abstract. The thickness of thin liquid films is of great interest to industrial processes and life science. However, there are not appropriate quantitative experimental tools for an adequate study of film evolution in case of not-ideal conditions. Here, we show the application of a holographic system for the evaluation of the 3D topography and thickness of evolving protein films. We use a custom holographic microscope that combines quantitative phase imaging with materials engineering. This technique offers an unprecedented level of details and we anticipate that it will promote a deeper understanding of the underlying physics of thin film dynamics.

1 Introduction

Thin liquid films flows are of fundamental importance in numerous areas of engineering, geophysics and biophysics [1, 2]. These films display unusual dynamic behaviours in drainage and measurement of the thin fluid film thickness evolution during drainage is key to understand such phenomena [3, 4]. In turn, such knowledge can be utilized to design new materials, to guide technologies involving bubbles and to control biological functions involving bubbles and bubbles rupture. Nevertheless, there are not appropriate quantitative experimental tools for an adequate study of film evolution in case of not-ideal conditions.

Herein, we tackle this problem using a holographic technique that combines quantitative phase imaging with a custom setup designed to form and manipulate bubbles [5]. digital holography (DH) is often used in sensors and microscopes to assess small variations of concentration [6], cell mass profiling [7] or thickness in transparent samples below the diffraction limit [8, 9]. In particular, we use transmission digital holography (TDH) to directly observe protein films ranging from a thickness of H≈25μm down to the wavelength of a coherent laser source (H≈632.8nm). We report our findings on the local film thickness and surface structure of films determined by TDH and discuss the implications of using this optical technique for studying non-uniform transparent surfaces.

2 Materials and Methods

2.1 Materials

The protein films were based on an aqueous solution of egg white albumen (EWA). EWA came from whole eggs purchased at the local market. The percentage of solid content of EWA was measured before use. Only EWA with 14% solid content was used for our experiments. The solvent used to dilute our protein mixture was DPBS, provided by Sigma.

2.2 Experimental setup

The setup consists of an off-axis Mach-Zhender interferometer and engineered for in order to allow the formation of thin liquid films and to control the bubble growth, see Figure 1. The films were formed on top of a metal pipe with a side inlet that was connected to a syringe pump. The illumination source was a HeNe laser (λ = 632.8nm). Indeed, using a laser source to illuminate a thin-film could have some effects like dehydration. However, in most case, the laser power used for thin-film holographic recording is very weak (<0.5mw), therefor the dehydration caused by laser recording is negligible compared with the dehydration caused by the overall temperature and humidity of the environment.

In the Mach-Zehnder interferometer, the laser beam is divided in two parts by the first beam-splitter cube (BS_1). The resulting beams are referred as object and reference beam. The object beam illuminates the sample from the top and forms the image on the camera. On the contrary, the reference beam goes towards the camera without encountering the sample. The two beams are collected by a second beam-splitter cube (BS_2). The cube is slightly tilted so that the two beams overlap with a small angle. This controls the period of the interference fringes, which can be adjusted according to the sampling requirements. Since the protein films usually have a complex surface structure, the Fresnel Hologram recording system was

* Corresponding author: z.wang@isasi.cnr.it

applied. Lens L_1 and lens L_2 form a 4F system to collect light beams.

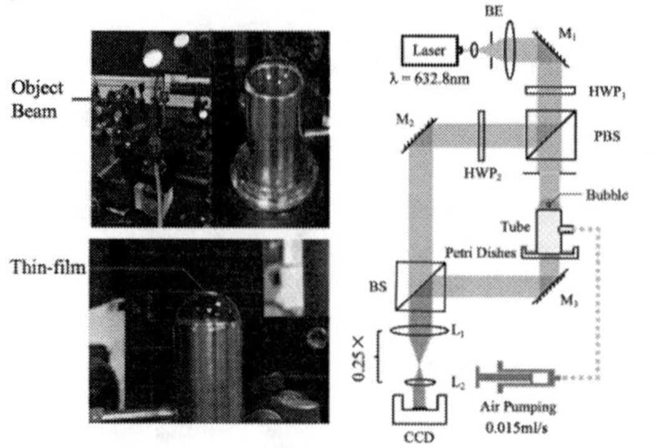

Fig. 2. Schematic of the transmission digital holography (TDH) setup that was used for this work. A laser source emits coherent single wavelength light into a beam expander (BE) and is then redirected by a mirror (M_1). The light passes through a halfway plate (HWP1) and is split into two beams by first beam-splitter cube (BS). The reference and object beams are collected by second bean-splitter cube (BS) and pass through two lenses (L_1 and L_2) before being collected by a CCD or CMOS for holographic numerical reconstruction processing.

3 Experimental results

We have studied the growth and drainage of protein liquid bubbles formed from diverse dilutions of EWA. We examined six different dilutions (5%, 20%, 40%, 60%, 80%, and 100%) and for each dilution we performed 5 experiments. Each bubble was grown following the same protocol. First, the bubble was created on top of the metal pipe, see Figure 1. Once a stable film was formed the laser shutter was open and the recording begun. At the same time, the syringe pump started inflating air through the inlet and the bubble was let grow for 30s. At this point the bubble was let drain and observed for around 3 minutes, see Figure 2.

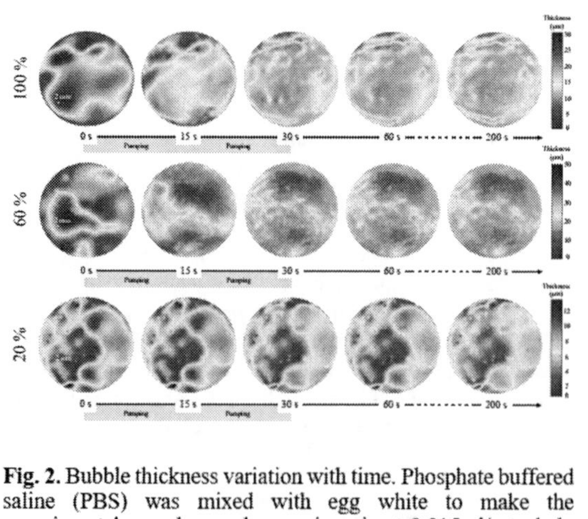

Fig. 2. Bubble thickness variation with time. Phosphate buffered saline (PBS) was mixed with egg white to make the experimental samples, and, pumping air at 0.015ml/s to help bubble film grow. Each set of experiments lasted 200 s, pumping air for the first 30 seconds, then stopped. The figure shows, from top to bottom, the thickness distribution of 100% egg white film, 60% egg white film and 20% egg white film.

In general, EWA bubbles showed a high level of stability. The drainage was mainly limited to the first seconds of the pumping phase, after which the system reaches an equilibrium. However, this equilibrium does not correspond to a homogeneous distribution of the proteins along the bubble's surface. Indeed, after 200s it is still possible to observe macro areas of protein aggregates all along the surface of the bubbles at every concentration.

4 Conclusions

In conclusion, we proposed a system that gathers all the features required to study evolving liquid thin film. First, the spatial resolution is limited only by the optics used. Second, it gives the full-field three-dimensional information of the sample, unlike techniques that use photomultipliers to have very fast measurement but only in one point. Third, DH does not require multiple exposures and can be matched with high-speed cameras to measure rapidly changing features, such as the rim of the hole formed by the rupture of the liquid film.

Finally, our study suggests that the use of EWA protein solutions to cast bubble films results in unique features that were not previously observed when detergents are used. Moreover, liquid films made from detergents are typically uniform and at a thickness range suitable for analysis by interferometry. On the other hand, our protein films do not show interference patterns owing to a complicated surface structure causing light scattering and are tens of microns thick, which would not result in interference patterns for visible wavelengths.

References

[1] P. Marmottant and S. Hilgenfeldt, Nature 423, 153–156 (2003);
[2] A. Salerno, E. Di Maio, S. Iannace, and P. A. Netti, J. Porous Mater. **19**, 181–188 (2012) ;
[3] L. Champougny, M. Roché, W. Drenckhan, and E. Rio, Soft Matter **12**, 5276–5284 (2016) ;
[4] D. Tammaro, R. Pasquino, M. M. Villone, G. D'Avino, V. Ferraro, E. Di Maio , A. Langella, N. Grizzuti, and P. L. Maffettone, Langmuir, **34**, 5646-5654 (2018);
[5] B. Mandracchia, Z. Wang, V. Ferraro, M. M. Villone, E. Di Maio, P. L. Maffettone, and P. Ferraro, Light Sci. Appl **8**, 20 (2019) ;
[6] F. Yesilkoy, R. A. Terborg, J. Pello, A. A. Belushkin, Y. Jahani, V. Pruneri, and H. Altug, Light Sci. Appl. 8, 17152 (2018) ;
[7] T. a Zangle and M. a Teitell, Nat. Methods **11**, 1221–1228 (2014) ;
[8] B. Mandracchia, O. Gennari, V. Marchesano, M. Paturzo, and P. Ferraro, J. Biophotonics **10**, 1163–1170 (2017) ;
[9] B. Mandracchia, O. Gennari, A. Bramanti, S. Grilli, and P. Ferraro, J. Biophotonics **11**, e201700332 (2018).

https://doi.org/10.1051/epjconf/201921513001

Femtosecond Laser Micromachining of Fabry-Pérot Interferometers for Magnetic Field Sensing

João M. Maia[1,2*], *Vitor A.* Amorim[1,2], *Duarte* Viveiros[1,2], and *P. V. S.* Marques[1,2]

[1]CAP – Centre for Applied Photonics, INESC TEC, Rua do Campo Alegre 4150-179, Porto, Portugal
[2]Department of Physics and Astronomy, Faculty of Sciences of University of Porto, Rua do Campo Alegre 4169-007, Porto, Portugal

Abstract. Fs-laser micromachining is a high precision fabrication technique that can be used to write novel three-dimensional structures, depending on the nature of light-matter interaction. In fused silica, the material modification can lead to (i) an increase of the refractive index around the focal volume, resulting in the formation of optical circuits, or (ii) an enhancement of the etch rate of the laser-affected zones relative to the pristine material, leading to a selective and anisotropic etching reaction that enables fabrication of microfluidic systems. Here, both effects are combined to fabricate a Fabry-Pérot interferometer, where optical waveguides and microfluidic channels are integrated monolithically in a fused silica chip. By filling the channel with a magnetic fluid whose refractive index changes with an external magnetic field, the device can be used as a magnetic field sensor. A linear sensitivity of -0.12 nm/mT is obtained in the 5.0±0.5 to 33.0±0.5 mT range, with the field being applied parallel to the light propagation direction.

1 Introduction

Ferrofluids are stable colloidal suspensions of single-domain magnetic nanoparticles dispersed in an aqueous or nonpolar solvent. Under an external magnetic field, the nanoparticles start to agglomerate in a chain-like structure oriented with the field direction, thereby changing the fluid's optical properties: refractive index, birefringence, dichroism, field-dependent transmission [1, 2]. Taking advantage of this phenomenon, several fiber-based magnetic field sensors for measurement of field magnitude or orientation have been demonstrated [3-5].

So far, photonic devices based on ferrofluids have only been used in planar systems for light modulation applications [6]. Compared to the existent fiber-based sensors, planar devices offer the chance of simultaneous detection of the field strength and its direction, made available not only due the ferrofluid properties but also by the advances made in fs-laser micromachining [7]. Ergo, this paper discusses the fabrication of Fabry-Pérot interferometers in fused silica by fs-laser micromachining and their response to magnetic field variations.

2 Experimental

The sensor's operation is based on wave interference, and since the contrast between the ferrofluid and fused silica is small, two-wave interference is enough to describe the process. The sensor is also designed as a reflective structure to maximize visibility. The reflected intensity (I_r) is given by:

$$I_r = I_1 + I_2 + 2\sqrt{I_1 I_2}\cos\delta \qquad (1)$$

where I_1 and I_2 are the intensities of the light beams reflected at the first and second interfaces, respectively, and have an impact on visibility. The condition for constructive interference is determined by the phase difference (δ) between both waves which relies on the cavity length and the ferrofluid refractive index. The latter further depends on the direction and magnitude of the magnetic field. As a result, the proposed device can be used as a magnetic field sensor.

2.1. Fabrication

A fiber amplified fs-laser system, emitting a second harmonic beam at 515 nm with an approximate pulse duration of 250 fs at 500 kHz, was used to write the interferometer. The laser beam was focused inside a fused silica substrate (Suprasil 1) with a 0.55 numerical aperture aspherical lens mounted in a vertical piezostage. The substrate was mounted on X-Y air-bearing linear stages and translated in relation to the beam focus. The optical waveguide and microfluidic channel were written sequentially to guarantee perfect alignment between both.

A 1.2-cm long waveguide was written 55 µm below the silica surface at a speed of 50 µm/s, pulse energy of 75 nJ and with the polarisation of the laser beam parallel to the writing direction. These parameters yield low scattering, which affects the Fabry-Pérot's signal-to-noise ratio. The microfluidic channel was written from bottom-to-top by stacking multiple laser tracks with a horizontal and vertical spacing of 2 and 3 µm respectively, at a speed of 500 µm/s, pulse energy of 80 nJ and with the beam's polarisation perpendicular to the scanning direction. In these conditions, the channel has low surface roughness.

* Corresponding author: joao.m.maia@inesctec.pt

Etching was done in an ultrasonic bath by immersing the substrate in a 10% hydrofluoric acid solution for 240 minutes. The channel has a final height of 90 μm and a tapered cross-section; the cavity length is 35 μm. The waveguide and the channel wall are distanced by 8 μm.

The substrate's facet was polished to improve light coupling. To inject the ferrofluid and to seal the device, a PDMS layer with two inlets/outlets was permanently bonded to the silica substrate. Sealing the device is necessary as the ferrofluid rapidly evaporates. A water-based ferrofluid (EMG507, Ferrotec Inc.), composed by Fe_3O_4 magnetic nanoparticles with average diameter of 10 nm in a volume concentration of 0.4-1.1% was used.

Fig. 1. Top-view of the optical waveguide orthogonally aligned with the microfluidic channel.

3 Results and Discussion

To measure the magnetic field, light from a broadband source is injected into an optical circulator and coupled into a single-mode fiber (SMF-28) that is butt-coupled to the waveguide. Light reflected in the Fabry-Pérot is transmitted back into the circulator, before being detected by an optical spectrum analyser. The substrate was mounted on a 5-axis stage for a precise alignment of the optical fiber with the waveguide. Index matching was used to minimize Fresnel reflections at the facet. A permanent magnet placed close to the substrate generates a magnetic field parallel to the light propagation direction. Its magnitude can be controlled by changing the distance between the magnet and the substrate. The measurements were made right after the ferrofluid was injected to prevent the nanoparticles from settling in the channel bottom. The temperature remained at 21.0±0.5°C and, in turn, thermally-induced index variations can be neglected.

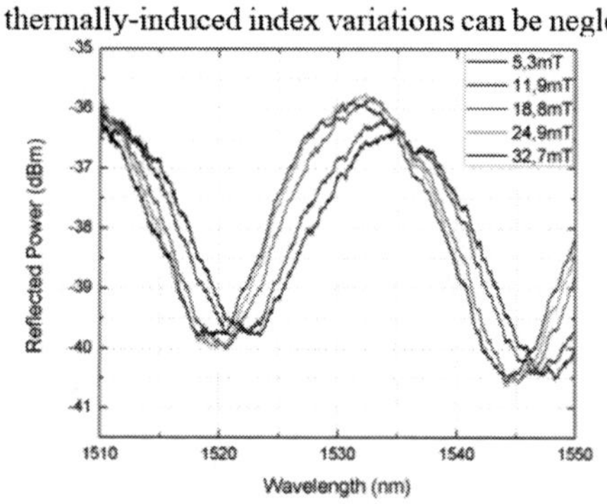

Fig. 2. Reflected spectrum at different field magnitudes.

As can be observed in Fig. 2, the interference spectrum shows a blue-shift when the field strength increases, indicating that the fluid's refractive index is decreasing. This is contradictory to what is expected [1-4] and can be explained by the fact that the nanoparticles are attracted towards the magnet and, consequently, they exit the cavity leading to a smaller concentration of nanoparticles and a decreasing refractive index. The low visibility of around 4 dB is due to the high optical absorption coefficient of the ferrofluid (≈20 dB per 100 μm). After smoothing the measured spectra, we retrieved the peak wavelength for each field magnitude (Fig. 3). A linear sensitivity of -0.12 nm/mT was obtained across the entire measured range. The use of ferrofluids with higher concentration of magnetic nanoparticles can enhance the sensitivity [3, 4].

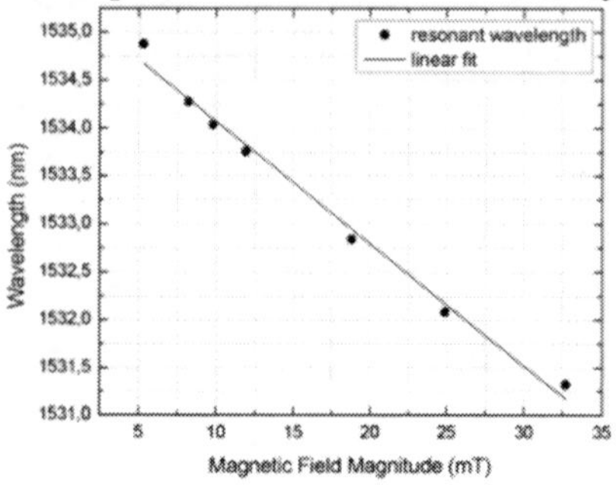

Fig. 3. Peak wavelength as a function of field magnitude.

4 Conclusion

Due to the tunable refractive index property of ferrofluids, an optofluidic sensor for magnetic field detection was fabricated by fs-laser micromachining. Preliminary results show that the sensor responds to field variations with a sensitivity of -0.12 nm/mT from 5.0±0.5 to 33.0±0.5 mT. Due to the 3D capabilities of fs-laser micromachining, other functionalities, enabled by the ferrofluid properties, can be envisioned: measurement (i) of field orientation by writing two other waveguides along the remaining spatial directions, or (ii) of time-varying fields through spectral filtering with Bragg gratings.

This work was supported by Fundação para a Ciência e Tecnologia through grant no. SFRH/BD/133095/2017 and by project "NanoSTIMA: Macro-to-Nano Human Sensing: Towards Integrated Multimodal Health Monitoring and Analytics/NORTE-01-0145-FEDER-000016", financed by the North Portugal Regional Operational Programme (NORTE 2020), under the PORTUGAL 2020 Partnership Agreement, and through the European Regional Development Fund (ERDF).

References

1. S.Y. Yang, J.J. Chieh, H.E. Horng, C.-Y. Hong, H.C. Hang, Appl. Phys. Lett. **84**, 25 (2004)
2. Z. Di, X. Chen, S. Pu, X. Hu, Y. Xia, Appl. Phys. Lett. **89**, 21 (2006)
3. R. Blue, A. Duduś, D. Uttamchandani, IEEE J. Sel. Topics Quantum Elceton. **22**, 2 (2016)
4. D. Homa, G. Pickrell, Sensors **14**, 3 (2014)
5. Z. Jiang, J. Dong, S. Hu, Y. Zhang, Y. Chen, Y. Luo, W. Zhu, W. Qiu, H. Lu, H. Guan, Y. Zhong, J. Yu, J. Zhang, Z. Chen, Opt. Lett. **43**, 19 (2018)
6. Y. Gu, F. Bragheri, G. Valentino, K. Morris, N. Bellini, R. Osellame, Appl. Opt. **54**, 6 (2015)
7. K. Sugioka, J. Xu, D. Wu, Y. Hanada, Z. Wang, Y. Cheng, K. Midorikawa, Lab Chip **14**, 18 (2014)

 https://doi.org/10.1051/epjconf/201921513002

Laser printed flat lenses for optofluidics

Airidas Žukauskas[1,*], *Andreas* R. Stilling-Andersen[1], *Xiaolong* Zhu[1], and *Anders* Kristensen[1]

[1]DTU Health Tech, Technical University of Denmark, DK-2800 Kongens Lyngby, Denmark

Abstract. Conventional three-dimensional optics requires curvature to control the wave front of light thus making it difficult to reduce the size of the optical systems. Furthermore, for correction of optical aberrations, complex optical systems comprising more than one lens are used. This adds additional bulk, mass and complexity to the optical systems. Recent development in diffractive optics has enabled new thin lightweight optical elements such as metalenses. We introduce resonant laser printing technique as a flexible photo-thermal technology for metalens fabrication with the ability to control the light with microscale precision. Our laser printed metalenses can be integrated in bio-sensors, bio-imaging systems, and optofluidical devices.

1 Introduction

Optical metasurfaces are arrays of sub-wavelength sized elements - meta-atoms - enabling precise control of phase, amplitude and polarization of light [1]. These metasurfaces are made from metallic (e.g. aluminum [2]) or high-index (e.g. silicon [3]) nano or micro sized arrays of meta-atoms that gives the surface unique optical features. These properties are defined by the material composition, the geometry and the orientation. These arrays are usually fabricated by Electron-beam [4] or Deep UltraViolet [5] lithography techniques which are multi-step, time consuming and nonflexible, making the flats lenses expensive and difficult to fabricate.

We introduce resonant laser printing (RLP) as a technique for fast and precise metasurface modification [6–9]. Nanosecond laser pulses are used for this process as the meta-atoms absorb the laser pulses and converts the energy to heat, which raises the temperature of the meta-atoms and causes them to melt and solidify in a short time period. This process changes the geometry of the structure and results in new optical properties of the metasurface. By controlling the laser pulse energy, it is possible to precisely tune the dewetting process.

We created metasurfaces with double resonances: one resonance in the visible - at the wavelength of the laser, we used for writing - to improve laser writing efficiency by decreasing the laser power required to modify the meta-atoms. The other resonance is the transmission resonance at the infra-red operating wavelength - that is very important for our flat lenses to work. Laser induced photo-thermal melting of the meta-atoms shifts the resonance at the infra-red operation wavelength, when the geometry of the meta-atom changes, it causes the resonance in transmission to blue shift, thus making the area more transparent.

Combining pre-fabricated, large area metasurface templates and the resonant laser printing technique, we get a low-cost and flexible fabrication technique of the metalenses that can be used for biological sensing, biological imaging, and optical trapping.

2 Results and discussion

In this work, we used pre-fabricated silicon based metasurface templates with double resonances at 532 nm and at 1064 nm. These templates were used to resonant laser print binary amplitude based Fresnel zone plates. We designed our metalens to work at a wavelength of 1064 nm (Fig. 1) with the focus length: 100 μm, number of rings/zones: 32, numerical aperture (NA): 0.7.

The laser written metalens was characterized by a custom built setup (Fig. 2), and we measured a focus length of 113 μm, beam FWHM of 2.29 μm at focus, the efficiency of the lens: 16.98%, and the lens diameter was 122 μm. Because of its small size, it can be integrated in optofluidics devices [10].

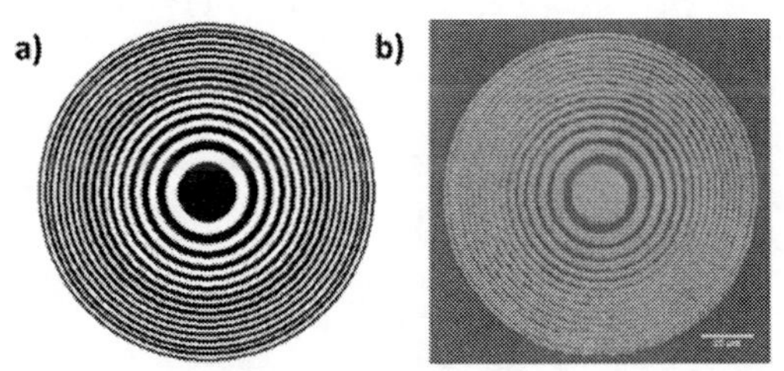

Figure 1. a) Our metalens design with the focus length: 100 μm, ring number: 32, numerical aperture (NA): 0.7, 305×305 pixels image. b) Transmission optical microscope image of the laser written metalens onto silicon based metalens templates. The scale bar is 20 μm.

*e-mail: aiuk@dtu.dk

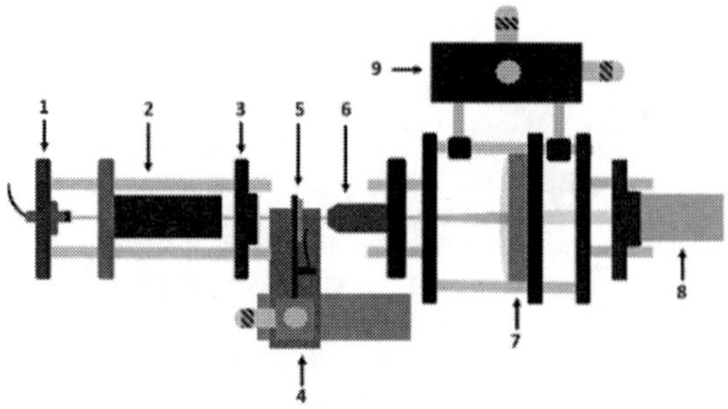

Figure 2. Schematics of out metalens characterization setup. 1) 1064 nm diode laser, 2) Glan-Thompson polarizer, 3) $\lambda/4$ plate, 4) motorized XYZ and rotation stage with sample holder, 5) sample, 6) 20× objective, 7) f = 150 mm lens, 8) CCD camera, 9) XYZ stage.

3 Conclusions

In conclusion, we demonstrated a powerful resonant laser writing technique for ultra-thin micro sized metalenses fabrication. These metalenses can be used for bio-imaging or for bio-sensors. Because of the very small size, it is possible to integrate these metalenses into optofluidic devices. Using prefabricated templates and laser fabrication, the cost of metalens is low, which provides a path for mass-production and ready-for-applications processes.

4 Acknowledgements

This work was supported by the European Commission project CHROMAVISION (Grant Agreement No. 665233). X.Z. is funded by VILLUM Experiment (grant No. 17400) supported by VILLUM FONDEN.

References

[1] N. Yu, F. Capasso, Nature Materials **13**, 139 (2014)
[2] E. Hjlund-Nielsen, X. Zhu, M.S. Carstensen, M.K. Sørensen, C. Vannahme, N.A. Mortensen, A. Kristensen, Optics Express **23**, 28829 (2015)
[3] Z. Zhou, J. Li, R. Su, B. Yao, H. Fang, K. Li, L. Zhou, J. Liu, D. Stellinga, C.P. Reardon et al., Acs Photonics **4**, 544 (2017)
[4] F. Yue, D. Wen, J. Xin, B. Gerardot, J. Li, X. Chen, p. 17 (2016)
[5] M. Iwanaga, 2016 Progress in Electromagnetics Research Symposium, Piers 2016 - Proceedings pp. 7734815, 1857–1861 (2016)
[6] X. Zhu, C. Vannahme, E. Højlund-Nielsen, N.A. Mortensen, A. Kristensen, Nature Nanotechnology **11**, 325 (2016)
[7] X. Zhu, W. Yan, U. Levy, N.A. Mortensen, A. Kristensen, Science Advances **3**, e1602487, e1602487 (2017)
[8] X. Zhu, M. Keshavarz Hedayati, S. Raza, U. Levy, N.A. Mortensen, A. Kristensen, Nano Today **19**, 7 (2018)
[9] M.S. Carstensen, X. Zhu, O.E. Iyore, N.A. Mortensen, U. Levy, A. Kristensen, A C S Photonics **5**, 1665 (2018)
[10] A. Žukauskas, X. Zhu, A. Kristensen (2019), in progress

https://doi.org/10.1051/epjconf/201921513003

Bi-phase emulsion droplets as dynamic fluid optical systems

Sara Nagelberg[1], *Amy Goodling*[2], *Kaushikaram Subramanian*[3], *George Barbastathis*[1], *Moritz Kreysing*[3], *Tim Swager*[4], *Lauren Zarzar*[2], and *Mathias Kolle*[1,*]

[1] Department of Mechanical Engineering, Massachusetts Institute of Technology, Cambridge, MA, USA.
[2] Department of Materials Science and Engineering, The Pennsylvania State University, University Park, PA, USA.
[3] Max Planck Institute of Molecular Cell Biology and Genetics, Dresden, Germany.
[4] Department of Chemical Engineering, Massachusetts Institute of Technology, Cambridge, MA, USA.

Abstract. Micro-scale optical components play a critical role in many applications, in particular when these components are capable of dynamically responding to different stimuli with a controlled variation of their optical behavior. Here, we discuss the potential of micro-scale bi-phase emulsion droplets as a material platform for dynamic fluid optical components. Such droplets act as liquid compound micro-lenses with dynamically tunable focal lengths. They can be reconfigured to focus or scatter light and form images. In addition, we discuss how these droplets can be used to create iridescent structural color with large angular spectral separation. Experimental demonstrations of the emulsion droplet optics are complemented by theoretical analysis and wave-optical modelling. Finally, we provide evidence of the droplets utility as fluidic optical elements in potential application scenarios.

Bi-phase emulsion droplets with reconfigurable internal interface

Micrometre-scale optical elements have contributed significantly to the miniaturization of optical devices and instrumentation [1-3]. To this end, optofluidic devices using dynamic fluid lens materials represent an ideal platform to create versatile, reconfigurable, refractive optical components [4]. Droplets smaller than the capillary length, wherein surface tension is the dominant force, create curved interfaces between fluid volumes [5]. These curved interfaces can be used to reflect and refract light forming tunable fluid micro-lenses, as individual components or arranged in arrays, for applications in miniaturized optical components and optics-based biosensing devices [8, 9]. Here, we present a strategy to form fluidic tunable compound micro-lenses from stable hydrocarbon and fluorocarbon emulsion droplets in aqueous media [10]. We use combinations of transparent fluorocarbon fluids, such as perfluorohexane or Fluorinert FC-770 (n = 1.27), and hydrocarbons, such as heptane (n = 1.39) or hexane (n = 1.38). These bi-phase emulsion droplets can be dynamically reconfigured between double emulsion and Janus (two-sided) morphologies through application of external stimuli, including chemical triggers, light, and heat, which makes them very promising as highly tunable compound lenses (Fig. 1).

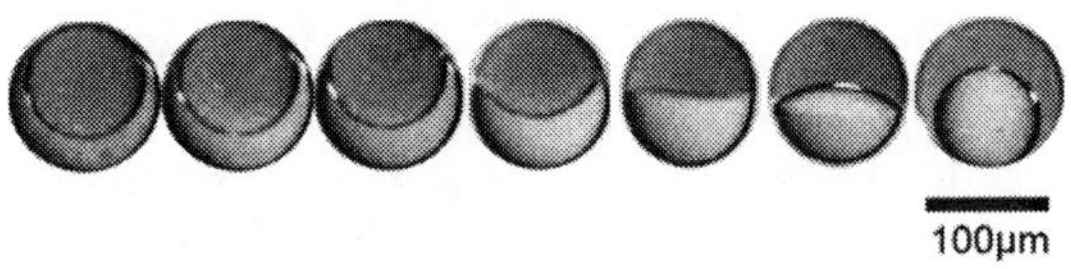

Fig. 1. Side view optical micrographs of droplets composed of FC-770 (grey) and heptane (red) with varying internal interface curvature. The red color of the heptane phase results from the incorporation of a Rhodamine-based dye.

Bi-phase droplets as tunable lenses

Due to their curved surfaces and interfaces and the refractive index contrast of the droplet constituents and the aqueous medium the droplets can function as lenses. Surfactants can be used to modify the relative interfacial tensions between the droplet phases and the surrounding medium to shape the droplet internal interfaces and thereby alter their lensing characteristics, such as the focal length (Fig. 2). In the presentation, we discuss potential application scenarios for such reconfigurable lenses in displays and 3D micro-imaging devices.

* Corresponding author: mkolle@mit.edu

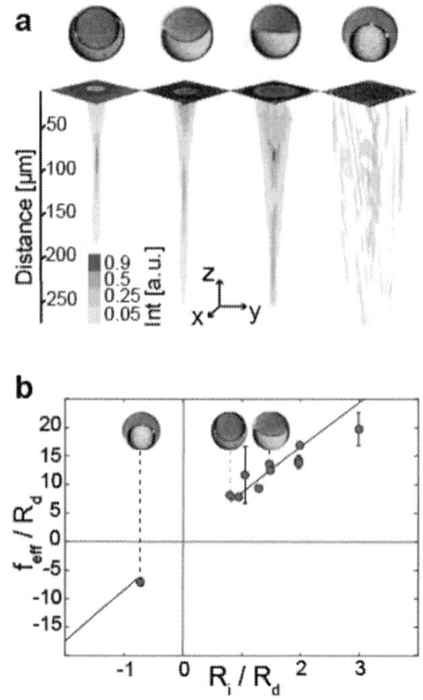

Fig. 2. a) Iso-surfaces of the reconstructed light fields behind the bi-phase droplets for different morphologies. b) Effective focal length as a function of internal radius of curvature R_i, normalized by the droplet diameter R_d.

Color generation in bi-phase emulsion droplets

In addition to light focusing, imaging, and beam steering, these fluid droplets can also be employed to create iridescent structural colors by cascaded total internal reflection and interference. Colors can generally result from many physical phenomena, including spectrally selective light absorption by pigments and dyes, material-specific optical dispersion and light interference in micro- and nanometre-scale periodic structures. Scattering, diffraction and interference mechanisms are particularly relevant to spherical droplets, which contribute to atmospheric phenomena such as glories, coronas and rainbows [11, 12]. The bi-phase emulsion droplets support a previously unrecognized mechanism for creating iridescent structural color with large angular spectral separation. The effect relies on light travelling along different trajectories of total internal reflection at a concave optical interface, causing interfere that results in brilliant color patterns. Each droplet acts as a spatio-spectral filter of incident light independent of other droplets in its surroundings (Fig. 3a). The color separation effect strongly depends on droplet size and morphology providing several different handles for controlling the phenomenon (Fig. 3b). A detailed characterization of the droplet colors, accompanied by theoretical modeling will be presented

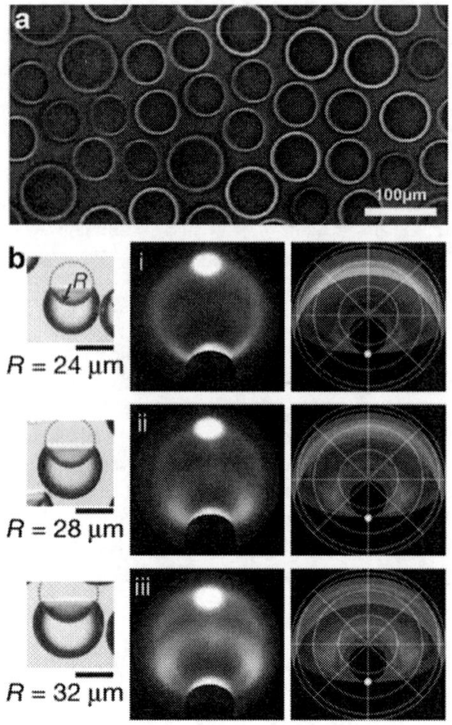

Fig. 3. a) Poly-disperse bi-phase emulsion droplets showing a size dependence of their color. b) Comparison of experimental iridescence maps of index-matched Janus droplets with the predicted three-dimensional calculation for various sizes.

References

1. Erdmann, L. & Gabriel, K. J., Appl. Opt. 40, 5592–5599 (2001).
2. Martinez-Cuenca, R., Saavedra, G., Martinez-Corral, M. & Javidi, B., J. Disp. Technol. 1, 321–327 (2005).
3. Xiao, X., Javidi, B., Martinez-Corral, M. & Stern, A., Appl. Opt. 52, 546–560 (2013).
4. Hawkins, A. R. & Schmidt, H., Handbook of Optofluidics (CRC Press, 2010).
5. Ren, H., Xu, S. & Wu, S.-T., Opt. Commun. 283, 3255–3258 (2010).
6. Kim, J., Singh, N. & Lyon, L. A., Angew. Chem. Int. Ed. 45, 1446–1449 (2006).
7. 41. Kuswandi, B., Nuriman, Huskens, J. & Verboom, W., Anal. Chim. Acta 601, 141–155 (2007).
8. Zarzar, L. D. et al., Nature 518, 520–524 (2015).
9. Hulst, H.C. *Light Scattering by Small Particles* (Courier Corporation, Mineola, 1981).
10. Nussenzveig, H. M., *J. Math. Phys.* **10**, 125–176 (1969).

Tunable optical lattices in the near-field of a few-mode nanophotonic waveguide

Christophe Pin[1,2,#], *Jean-Baptiste* Jager[2], *Manon* Tardif [2], *Emmanuel* Picard[2], *Emmanuel* Hadji[2], *Frédérique* de Fornel[1], and *Benoît* Cluzel[1,*]

[1]Groupe Optique de Champ Proche – L.R.C. SiNOPTIQ du C.E.A. n°DSM-08-36, Laboratoire Interdisciplinaire Carnot de Bourgogne U.M.R. C.N.R.S. 6303, Université de Bourgogne-Franche Comté, 9. Av. A. Savary, 21078 Dijon, France
[2]Université Grenoble Alpes, C.E.A., I.N.A.C., PHELIQS, SINAPS, F-38000 Grenoble, France
[#] *Current address:* RIES, Hokkaido University, Kita 20-jo Nishi 10-chome, Kita-ku, Sapporo, Hokkaido, 001-0020, Japan

Abstract. Due to the action of the scattering force, particles that are optically trapped at the surface of a waveguide are propelled in the direction of the light propagation. In this work, we demonstrate an original approach for creating tunable periodic arrays of optical traps along a few-mode silicon nanophotonic waveguide. We show how the near-field optical forces at the surface of the waveguide are periodically modulated when two guided modes with different propagation constants are simultaneously excited. The phenomenon is used to achieve stable trapping of a large number of dielectric particles or bacteria along a single waveguide. By controlling the light coupling conditions and the laser wavelength, we investigate several techniques for manipulating the trapped particles. Especially, we demonstrate that the period of the optical lattice can be finely tuned by adjusting the laser wavelength. This effect can be used to control the trap positions, and thus transport the trapped particles in both directions along the waveguide.

1 Introduction

Photonic waveguides are commonly used as on-chip optical conveyor belts in on-chip optical trapping and manipulation experiments. Kawata and Tani [1] first demonstrated in 1996 that particles in liquid solution can be trapped and propelled under the action of the optical force in the near-field of a photonic waveguide. Since then, various types of waveguides have been studied for optical trapping and transportation of particles: ridge [1-5], slot [6], photonic crystal [7], plasmonic [8], hollow-core waveguides [9], as well as micro- and nanofibres [10]. Waveguides have been mostly used for efficiently transporting micro- and even nanoparticles [6], as well as cells and bacteria in liquid solutions [3,4,9]. However, stable optical trapping of particles at a constant position along the waveguide is a more challenging task. In previous works, it required the use of a photonic or plasmonic resonator [11-20], or the presence of a standing wave along the waveguide, formed by two counter-propagating waves [21,22]. Some recent works suggested that the simultaneous excitation of two modes with different propagation constants could provide an alternative approach [2,23,24].

In this work, we investigate the optical force distribution in the near-field of a few-mode silicon waveguide. Numerical simulation and optical trapping experiments are performed in order to study the periodic modulation of optical forces resulting from the simultaneous excitation of different pairs of guided modes [25]. Especially, we achieve creating near-field optical lattices that enable stable optical trapping of dielectric particles and bacteria. We also report on the use of such near-field optical lattices in order to handle in various ways the trapped particles [25].

2 Results and discussion

2.1 Few-mode silicon waveguides

Nanophotonic waveguides with rectangular cross-sections were fabricated on a Silicon-On-Insulator chip using standard e-beam lithography process. The light from a tunable laser source emitting at telecom wavelengths was coupled into 5-µm-large waveguides using a lensed optical fibre. The position of the fibre relative to the waveguide entrance and the polarisation of the incident laser light were then used to control the excitation of the different guided modes in a few-mode nanophotonic waveguide on the chip.

In the case of a waveguide with a 510 x 248 nm² cross-section, three guided modes can be excited: two transverse-electric modes (TE0 and TE1) and one transverse-magnetic mode (TM0). When two guided modes are excited at the same time in the waveguide, the near-field optical forces are periodically modulated due to the difference of propagation constants. The period of the modulation follow the equation:

* Corresponding author: benoit.cluzel@u-bourgogne.fr

https://doi.org/10.1051/epjconf/201921514001

$$p_{ij} = 2\pi / |\beta^{(i)} - \beta^{(j)}| = \lambda_0 / |n_{eff}^{(i)} - n_{eff}^{(j)}| \quad (1)$$

where λ_0 refers to the free-space wavelength and $\beta^{(i)}$, $\beta^{(j)}$ (respectively $neff^{(i)}$, $neff^{(j)}$) to the propagation constants (respectively to the effective refractive indices) of the two guided modes. Because the propagation constant of each mode depends on the wavelength, we can evaluate the variation of the period of the optical force modulation based on numerical simulation results (Figure 1).

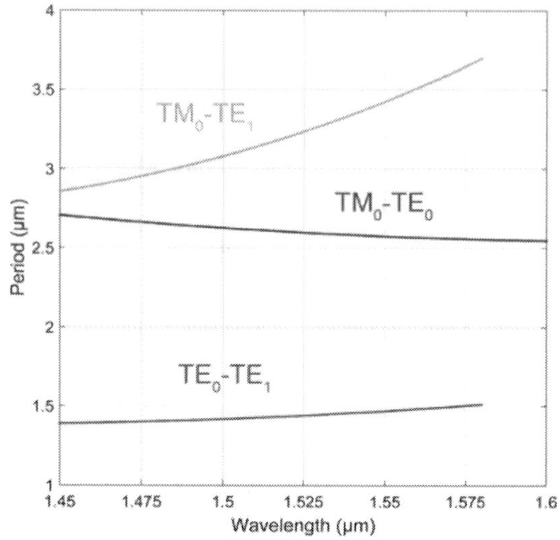

Fig. 1. Period of the optical force modulation as a function of the wavelength.

2.2 Near-field optical lattices

In some cases, the periodic modulation of the near-field optical forces can lead to the formation of a periodic array of optical traps along the surface of the waveguide. As shown in Figure 2, 1 µm and 500 nm fluorescent polystyrene beads as well as e. coli bacterium cells were stably trapped in such optical lattices when both transverse-electric (TE) and transverse-magnetic (TM) modes were excited in the waveguide. In Figure 2(a), 500 nm beads were trapped in staggered rows on both sides of the waveguide with a period of 1.46 µm. In Figure 2(b), a periodic array of 1 µm beads with a period of 3.26 µm was formed above the waveguide. Those experimental results are in good agreement with numerical simulation results. Indeed, it was verified that the simultaneous excitation of both the fundamental TE and TM modes (TE_0 and TM_0, respectively) is at the origin of the one-row periodic array of traps, while the simultaneous excitation of both the fundamental and the first order TE modes (TE_0 and TE_1, respectively) is at the origin of the particles being trapped in staggered rows.

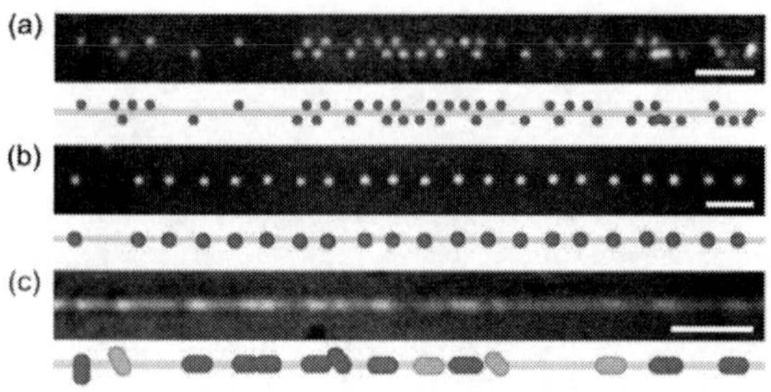

Fig. 2. Periodic arrangement of 500 nm polystyrene beads trapped in staggered rows. (b) Periodic arrangement of 1µm polystyrene beads above the waveguide. (c) Stable trapping of several e.coli bacteria above a single waveguide.

2.3 On-chip optical manipulation

The near-field optical lattices described above enable not only stable trapping of a large number of particles but also the development of new on-chip optical manipulation techniques. By modifying the polarization of the incident laser light or the light coupling conditions at the entrance of the photonic chip, it was possible to switch either from one optical lattice configuration to another, or from the stable trapping to the optical propulsion of the trapped particles. Moreover, it is possible to dynamically control the period of the optical lattice by tuning the laser wavelength. The particle spacing can therefore be finely tuned, and the position of the trapped particles precisely controlled along the waveguide. As shown in Figure 3, we were able to transport the trapped particles in both directions along the waveguide By continuously varying the laser wavelength: in direction of the light propagation when increasing the wavelength (forward transportation), and in the opposite direction when decreasing the wavelength (backward transportation).

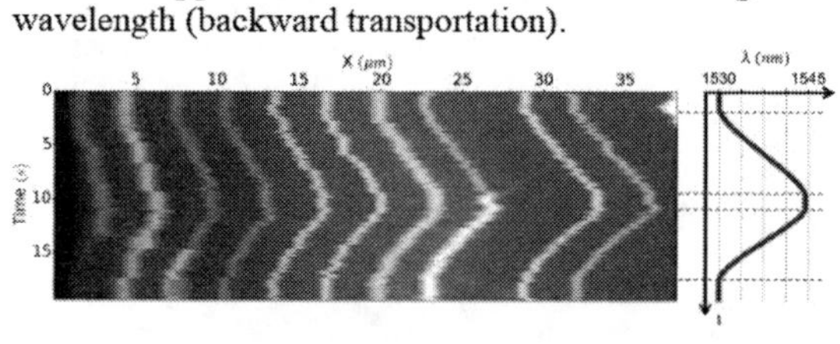

Fig. 3. Kymograph describing the continuous forward (to the right) and backward (to the left) transportation of 500 nm beads trapped in a TM_0-TE_1 lattice by continuously varying the wavelength.

3 Conclusion

In this work, it was demonstrated how a few-mode silicon waveguide can be used as a platform for creating one-dimensional near-field optical lattices in an optofluidic chip. It was shown that the simultaneous excitation of pairs of guided modes generates a periodic modulation of near-field optical forces along the waveguide. By selecting the adequate pairs of guided modes, we were able to stably trap a large number of particles and bacteria along a single waveguide. In addition, we showed how controlling the light coupling conditions and the laser wavelength makes it possible to finely tune the optical lattice properties. Taking advantage of the lattice tunability, we demonstrated several on-chip optical manipulation techniques for handling the trapped particles. The optical trapping and manipulation methods demonstrated in this work can be readily implemented in lab-on-a-chip devices and should pave the way for various optofluidic applications.

This work was performed in cooperation with the Labex ACTION program (contract ANR-11-LABX-0001-01), and with the support of the "Plateforme Technologique Amont" of Grenoble, funded through the CNRS Renatech network.

References

1. S. Kawata, T. Tani, Opt. Lett. **21**, 1768 (1996)
2. T. Tanaka, S. Yamamoto, Jpn. J. Appl. Phys. **41**, L260 (2002)
3. S. Gaugiran, S. Gétin, J.M. Fedeli, G. Colas, A. Fuchs, F. Chatelain, J. Dérouard, Opt. Express **13**, 6956 (2005)
4. B.S. Ahluwalia, P. McCourt, T. Huser, O.G. Hellesø, Opt. Express **18**, 21053 (2010)
5. P. Schein, D. O'Dell, D. Erickson, ACS Photon. **4**, 106 (2017)
6. A.H. Yang, S.D. Moore, B.S. Schmidt, M. Klug, M. Lipson, D. Erickson, Nature **457**, 71 (2009)
7. M.G. Scullion, Y. Arita, T.F. Krauss, K. Dholakia, Optica **2**, 816 (2015)
8. C. Yang, D. Pan, L. Tong, H. Xu, Nanoscale **8**, 19195 (2016)
9. S. Unterkofler, M.K. Garbos, T.G. Euser, P.S.J. Russell, J. Biophotonics **6**, 743 (2013)
10. G. Brambilla, G.S. Murugan, J.S. Wilkinson, D.J. Richardson, Opt. Lett. **32**, 3041 (2007)
11. S. Mandal, X. Serey, D. Erickson, Nano Lett. **10**, 99 (2009)
12. C. Renaut, J. Dellinger, B. Cluzel, T. Honegger, D. Peyrade, E. Picard, F. de Fornel, E. Hadji, Appl. Phys. Lett. **100**, 101103 (2012)
13. C. Renaut, B. Cluzel, J. Dellinger, L. Lalouat, E. Picard, D. Peyrade, E. Hadji, F. de Fornel, Sci. Rep. **3**, 2290 (2013)
14. S. Lin, K.B. Crozier, ACS Nano **7**, 1725 (2013)
15. C. Pin, B. Cluzel, C. Renaut, D. Peyrade, E. Picard, E. Hadji, F. de Fornel, Appl. Phys. Lett. **105**, 171108 (2014)
16. C. Pin, B. Cluzel, C. Renaut, E. Picard, D. Peyrade, E. Hadji, F. de Fornel, ACS Photon. **2**, 1410 (2015)
17. C. Pin, C. Renaut, M. Tardif, J.-B. Jager, E. Delamadeleine, E. Picard, D. Peyrade, E. Hadji, F. de Fornel, B. Cluzel, Proc. SPIE **10252**, 1025212 (2017)
18. P.T. Lin, H.Y. Chu, T.W. Lu, P.T. Lee, Lab. Chip **14**, 4647 (2014)
19. G. Magno, A. Ecarnot, C. Pin, V. Yam, P. Gogol, R. Mégy, B. Cluzel, B. Dagens, Opt. Lett. **41**, 3679 (2016)
20. A. Ecarnot, G. Magno, V. Yam, B. Dagens, Opt. Lett. **43**, 455 (2018)
21. M. Soltani, J. Lin, R.A. Forties, J.T. Inman, S.N. Saraf, R.M. Fulbright, M. Lipson, M.D. Wang, Nature Nanotech. **9**, 448 (2014)
22. J.E. Baker, R.P. Badman, M.D. Wang, WIREs Nanomed. Nanobiotechnol. **10**, e1477 (2018)
23. O.A. Schmidt, T.G. Euser, P.S.J. Russell, Opt. Express **21**, 29383 (2013)
24. T.H. Stievater, D.A. Kozak, M.W. Pruessner, R. Mahon, D. Park, W.S. Rabinovich, F.K. Fatemi, Opt. Mat. Express **6**, 3826 (2016)
25. C. Pin, J.-B. Jager, M. Tardif, E. Picard, E. Hadji, F. de Fornel, B. Cluzel, Lab. Chip **18**, 1750 (2018)

Real-time Measurement of the Single Nanoparticle Electrophoretic Mobility

Bohdan Yeroshenko[1,*], *Wouter* Wassing[1], *Allard P.* Mosk[1], and *Sanli* Faez[1,**]

[1]Utrecht University, Debye Institute for Nanomaterials Science, Utrecht, 3584 CS, Netherlands

Abstract. The electrophoretic mobility of a single nanoparticle depends on its surface charge and its environment. Thus the change of the mobility can reflect the change in its chemical and physical properties. We present a high-bandwidth method to measure the electrophoretic mobility, based on optical tweezers and electrophoresis. We envision studying of nanoscale chemical processes as a possible application of this method.

1 Introduction

A wide range of processes is happening on the surfaces of nanoparticles in both industrial chemistry and biology, such as catalysis and enzymatic activity. These processes are associated with changes to the particle properties such as surface charge, polarizability and hydrodynamic properties. One way to study these changes is by studying the electrophoretic mobility; by confining the particle in space by optical tweezers and applying an external oscillating electric field. It was shown that single charge transfer is measurable with this method in a non-polar solvent [1]. In other reports, electrophoresis was studied on single particles using digital signal processing [1][2][3]. We extend those methods to the real time measurement of the mobility and thus extract the charge of a nanoparticle.

2 Motion of a nanoparticle in an optical trap in an oscillating electric field

Diffusion of a single nanoparticle in a liquid can be described as a random walk, if we consider time scales bigger than those associated with inertial effects, $t_s > m/\gamma_0$. Then the motion of particle can be described as a Brownian noise:

$$S_{BN} = \frac{D/2\pi^2}{f^2}$$

where D is a diffusion constant and f is frequency. The description of the forces acting on the nanoparticle in the optical tweezers vary depending on the size of the particle in relation to the wavelength of the trapping beam, but can be split into the optical scattering force, that pushes the particle in the direction of the light propagation, and the optical gradient force that pushes the particle in or out of the highest intensity of the light. The force can be accurately approximated as a harmonic potential. The power spectrum of the motion will then be:

$$S_{OT} = \frac{D/2\pi^2}{f^2 - f_c^2}$$

where f_c is a corner frequency. This means that the optical trap effectively attenuates the low frequency motion, at frequencies smaller than the corner frequency of the trap, and does not affect the motion at higher frequencies. The fast displacements can thus be considered the same as that of a freely diffusing nanoparticle. When the external field is applied to a particle in a solvent, it will start to move in the field with a proportional drift velocity. The drift velocity is related to the external field and the electrophoretic mobility as

$$v_{ep} = \mu E.$$

where μ is an electrophoretic mobility and E is magnitude of the external electric field. The electrophoretic mobility depends on properties of the media the nanoparticle is in, such as its viscosity and dielectric constant, and the properties of the particle, its size and most importantly, for this application, the zeta potential. Zeta potential, ζ, is the electric potential at the slip plane, a plane in the electric double layer around the particle, that denotes a first plane from the surface of the particle, where the ions are mobile [4].

$$\mu = \frac{\epsilon_0 \epsilon_r}{\gamma}\zeta$$

By measuring the amplitude of the oscillation in space at that frequency, we can immediately measure its mobility and thus zeta potential and thus we can calculate the charge state of the particle. In the case when the applied voltage is harmonic, and the period of the applied voltage longer than the relaxation time of the charges around the particle, the motion at the applied frequency can be described as:

$$S_{EP} = \frac{|E^2|}{f^2}\delta(f_{ext}) * |\mathcal{F}\{\mu(t)\}|^2$$

where f_{ext} is a frequency of the applied field and $\mathcal{F}\{\mu(t)\}$ is a Fourier Transform of the mobility change in time. Thus the change of the mobility will be present in the power spectrum of the motion of the particle as a broadening of the peak around the applied frequency, see Fig.1.

*e-mail: b.yeroshenko@uu.nl
**e-mail: s.faez@uu.nl

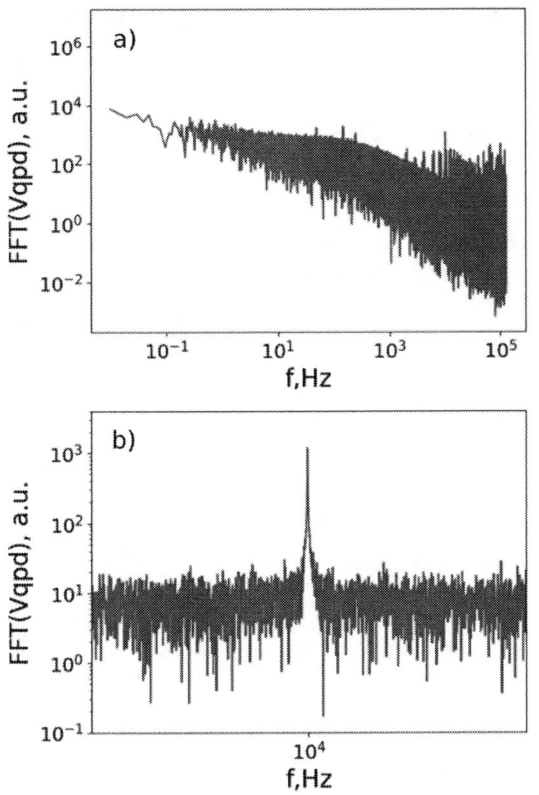

Figure 1. Fast Fourier Transform of the voltage readings of the quadrant photodiode (QPD) corresponding to the motion of the 490 nm TiO_2 nanoparticle inside of the optical tweezer a) full spectrum b) around the frequency of the applied field with characteristic broadening of the peak

Thus by using a lock-in amplifier one can directly measure changes to the electrophoretic mobility. Time resolution of this measurement is limited only by the low pass filter properties of the lock-in amplifier. The noise of the signal is the Brownian noise of the particle around the applied frequency. And the signal to noise ration (SNR) can be increased by applying a larger electric field, up to a certain limit dictated by the chemical stability of the solution.

3 Methods

A 1064 nm laser is used for trapping, with usual power of the beam being around 300 mW. Detection is done by both a camera and a quadrant photodiode (QPD). The QPD allows fast signal sampling, with maximum at around 700 kHz. It allows to continuously monitor the total reflected signal, which in turn allows detection of a trapping event and the relative position of the particle at the same time. We use a polarized beam splitter and a quarter-wave plate for maximization of signal collection. We apply the electric field through microfabricated electrodes in the microfluidic cell. The signal at the actuation frequency is separated using a lock-in amplifier.

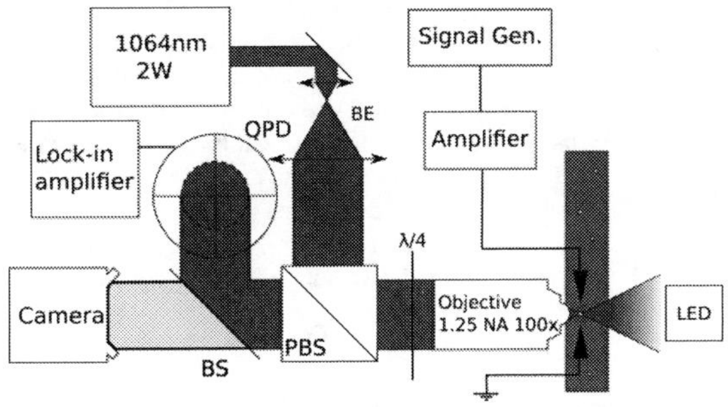

Figure 2. Schematic of the optical tweezer setup in the backscattering configuration

4 Discussion

Our experiment can measure the electrophoretic mobility in real time with time resolution of around 1 ms and is limited in duration by the trapping time which can be hours. This setup is able to provide both the precise measurement of the mobility of the particle, its change due to fluctuation of the charges in the electric double layer, and its change under external influence. With careful experimental design we expect that we will be able to measure polarizability, Stern potential and Zeta potential of a single nanoparticle. [4].

This research is supported by the Dutch Science Foundation (NWO).

References

[1] F. Beunis, Strubbe, F., K. Neyts, D. Petrov, Phys. Rev. Lett. **108** p.016101 (2012)

[2] O. Otto, C. Gutsche, F. Kremer, U. Keyser, Rev. Sci. Instrum. **79** p.023710 (2008)

[3] G.S. Roberts, T.A. Wood, W. Frith, P. Bartlett, J. Chem. Phys. **126** p.194503. (2007)

[4] S.S. Dukhin, B.W. Derjagin, *Electrophoresis* (Nauka, Moscow, 1976), pp. 74, 146

EPJ Web of Conferences **215**, 14003 (2019) https://doi.org/10.1051/epjconf/201921514003
EOS Optical Technologies

Design Of An Optofluidic Device For The Measurement Of The Elastic Modulus Of Deformable Particles

Massimiliano M. Villone[1,*], *Janine K.* Nunes[2], *Howard A.* Stone[2], and *Pier Luca* Maffettone[1]

[1]Department of Chemical, Materials, and Manufacturing Engineering, University of Naples Federico II
[2]Department of Mechanical and Aerospace Engineering, Princeton University

Abstract. Suspensions carrying deformable inclusions are ubiquitous in nature and applications. Hence, high-throughput characterization of the mechanical properties of soft particles is of great interest.
Recently, a non-invasive optofluidic technique has been developed for the measurement of the interfacial tension between two immiscible liquids [8]. We have adapted such technique to the case of soft solid beads, thus designing a non-invasive optofluidic device for the measurement of the mechanical properties of deformable particles from real-time optical imaging of their deformation.
The device consists of a cylindrical microfluidic channel with a cross-section reduction in which we make initially spherical soft beads flow suspended in a Newtonian carrier. By imaging the deformation of a particle in real time while it goes through the constriction, it is possible to get a measure of its elastic modulus through a theoretically derived-correlation. We provide both experimental and numerical validation of our device.

Suspensions carrying soft inclusions are frequent in nature and applications, e.g., biological cell suspensions. Their behavior is strongly influenced by the mechanical properties of the suspended particles, thus being able to measure these properties could be crucial to understand and control suspension flow. Moreover, biological cells can modify their deformability and functionality depending on their health state [1], so the measurement of their mechanical properties can be a powerful tool to discriminate whether they are healthy or not. In the last twenty years, several methods have appeared in the literature for the measurement of the mechanical properties of deformable particles, like atomic force microscopy-based techniques (see, e.g., [2] and the references therein), micropipette aspiration [3], compression between parallel plates [4], osmotic compression [5], and capillary micromechanics [6]. However, all the aforementioned techniques are quite time-consuming. Some of them are also 'invasive' for the investigated samples and the measured values of the mechanical properties can be influenced by the contact with the probe of the measuring instrument. In the very recent years, microfluidics is turning out to be suitable tool for the measurement of the mechanical properties of biological cells [7]. About a decade ago, Hudson and co-workers have developed a continuous-flow non-invasive microfluidic technique providing almost real-time measurement of the interfacial tension between a liquid droplet and an immiscible suspending liquid [8]. Their method offers considerable advantages in terms of device fabrication and operation ease, low costs, and high throughput. We propose an adaptation of such a technique to the measurement of the elastic modulus of deformable particles and we validate it through numerical simulations and experiments on homogeneous synthetic particles with sizes and elasticities in the range of interest for biological cells [1].

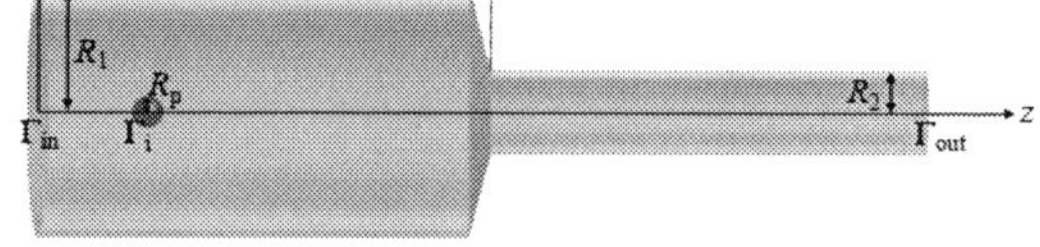

Figure 1. a) Schematic drawing of an elastic particle deformed into a prolate spheroid. b) Geometry of an initially spherical elastic bead suspended in a Newtonian liquid flowing through a device consisting of two coaxial cylindrical tubes.

In Fig. 1, the geometry of the device is depicted: an initially spherical elastic bead with radius R_p is suspended in a Newtonian liquid flowing through a cylindrical pipe with radius R_1. This channel undergoes an abrupt radial contraction to a narrow cylinder of radius R_2. While going through the constriction, the particle feels a time-dependent uniaxial extensional flow with extensional rate $\dot{\varepsilon}(t)$ that makes it deform into a prolate spheroid. If one monitors the transient response of the elastic bead to such flow and measures the deformation parameter $D(t) = (L(t) - B(t))/(L(t) + B(t))$ (see Fig. 1a) and its time derivative dD/dt, a plot of $\eta(15/8\dot{\varepsilon} - dD/dt)$ versus D (with η

*e-mail: massimiliano.villone@unina.it

the suspending liquid viscosity) should show a linear trend with a slope corresponding to the particle's elastic modulus G [9]. Therefore, such a flow measurement provides a means for measuring G in a continuous-flow process. This argument is conceptually analogous to that proposed by Hudson et al. for liquid droplets [8].

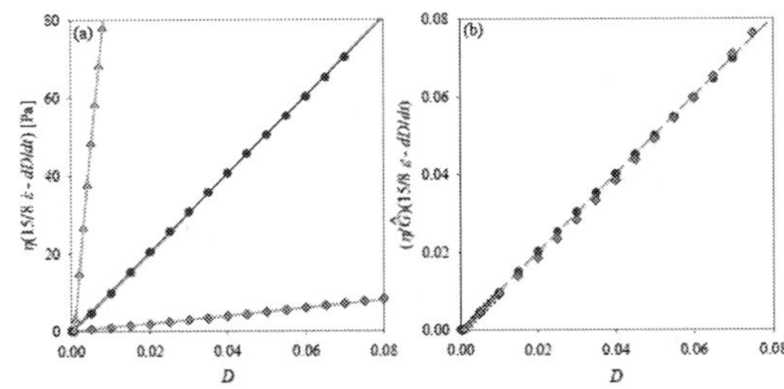

Figure 2. (a) Green diamonds, red circles, cyan triangles: Taylor plot of $\eta(15/8\dot{\varepsilon} - dD/dt)$ versus D for three different G-values. Green, red, cyan lines: linear fits of the three series of symbols (the slopes are equal to 100.71, 1009.5, 9998 Pa, respectively). (b) Taylor plot of $(\eta/\hat{G})(15/8\dot{\varepsilon} - dD/dt)$ versus D, with $\hat{G}$ the estimated value of the particle shear modulus, for the same data sets as in panel (a).

In order to validate our approach numerically, we have run simulations with beads having shear moduli of different orders of magnitude and verified that the aforementioned relation holds. Indeed, the data sets from numerical simulations appearing in Fig. 2a can be suitably fitted through linear functions with null intercept and a slope less than 1% different from the actual simulation input. In Fig. 2b, the sets of numerical results appearing in Fig. 2a are plotted each one scaled by the corresponding estimated value of the elastic modulus $\hat{G}$. As expected, the three series of data all collapse on the bisector of the first quadrant of the Cartesian plane. Experiments on PEG-DA microgel particles provide a further proof of the validity of the proposed device.

References

[1] Fraldi, M and Cugno, A and Deseri, Luca and Dayal, K and Pugno, N M., Journal of the Royal Society Interface **12**, 20150656 (2015)

[2] Tsukruk, Vladimir V and Singamaneni, Srikanth, *Scanning Probe Microscopy of Soft Matter: Fundamentals and Practices*, (John Wiley & Sons, New York, 2012)

[3] Guilak, Farshid and Tedrow, John R and Burgkart, Rainer, Biochemical and Biophysical Research Communications **269**, 781–786 (2000)

[4] Thoumine, Olivier and Ott, Albrecht and Cardoso, Olivier and Meister, Jean-Jacques, Journal of Biochemical and Biophysical Methods **39**, 47–62 (1999)

[5] Saunders, Brian R and Vincent, Brian, Advances in Colloid and Interface Science **80**, 1–25 (1999)

[6] Wyss, Hans M and Franke, Thomas and Mele, Elisa and Weitz, David A., Soft Matter **6**, 4550–4555 (2010)

[7] Mietke, Alexander and Otto, Oliver and Girardo, Salvatore and Rosendahl, Philipp and Taubenberger, Anna and Golfier, Stefan and Ulbricht, Elke and Aland, Sebastian and Guck, Jochen and Fischer-Friedrich, Elisabeth, Biophysical Journal **109**, 2023–2036 (2015)

[8] Hudson, Steven D and Cabral, Joao T and Goodrum Jr, William J and Beers, Kathryn L and Amis, Eric J, Applied Physics Letters **87**, 081905 (2005)

[9] Roscoe, R, Journal of Fluid Mechanics **28**, 273–293 (1967)

2D resolution improvement via 1D scanning Space-Time Digital Holography (STDH) in Optofluidics

Zhe Wang[1,2], *Vittorio Bianco*[1], *Yutong Cui*[2], *Melania Paturzo*[1*], *and Pietro Ferraro*[1]

[1]CNR-ISASI, Istituto di Scienze Applicate e Sistemi Intelligenti «E. Caianiello» del CNR, Via Campi Flegrei 34, 80078 Pozzuoli, Napoli, Italy.
[2]College of Applied Sciences, Beijing University of Technology, 100124 Beijing, China.

Abstract. Space-Time Digital Holography (STDH) exploits the object motion to record the hologram in a hybrid space-time domain. This representation adds new capabilities to conventional DH, such as unlimited extension of the Field of View (FoV) and tunable phase shifting. Here we show that STDH is able to improve the spatial resolution as well. Differently from other super-resolution approaches, stitching between holograms or their spectra is no longer required. Moreover, we introduce a new STDH modality to record and process hybrid space-time representations. This allows improving resolution with one single object scan, paving the way to the use of STDH for superresolution imaging onboard Lab on a Chip devices.

1 Introduction

Recently, we introduced a novel recording modality, named Space-Time Digital Holography (STDH), which exploits a projection of the DH fringes to map the hologram signature in a hybrid space-time domain [1-3]. The idea of STDH is to move complexity from the reconstruction to the acquisition process, by imposing a specific constraint on the scanning velocity that has to be matched to the frame rate of the sensor. Thus, a synthetic STDH can be created that possesses all the capabilities of a space-space DH [4,5], along with some added-value features. Above all, whenever the object motion is an intrinsic feature of the system, e.g. in the case of microfluidics, a STDH mapping all the samples flowing inside the channel during a long experiment can be generated and numerically refocused by one single propagation process. Thus, STDH has been shown to be a valuable tool for the analysis with extended FoV of biological samples in Lab-on-a-Chip (LoC) environment, and for high-throughput cell counting and velocimetry by exploiting the sample motion [3]. In this manuscript, we introduce a new STDH modality to record and process hybrid space-time representations. This allows improving resolution with one single object scan. Onion epidermal cells has been used as samples to verify this approach.

2 Methods

Take x-direction scanning as an example, when using an M×N area array CCD sensor for digital holographic recording of space-time scanning, M columns of pixels can be used for space-time hologram construction. Then, if K frame holograms are recorded, according to the basic space-time hologram composition method, we can build M frame space-time stacks:

$$I_i(k,y)=\begin{bmatrix} s_{1,0} & s_{2,0} & \cdots & s_{K,0} \\ s_{1,1} & s_{2,1} & \cdots & s_{K,1} \\ \vdots & & \ddots & \vdots \\ s_{1,N} & s_{2,N} & \cdots & s_{K,N} \end{bmatrix},\ (i=1,\dots,M) \tag{1}$$

where, $I_i(k,y)$ (i=1,…,M) represent space-time holograms constructed using different columns of pixels. Furthermore, above stacks can be thought of as a series of holograms containing continuously varying phase shift information, like shown below:

$$\begin{cases} I_0(k,y)=A(k,y)+2|E_0(k,y)||E_R(k,y)|\cos[\varphi(k,y)+\psi_0(k,y)+\Delta\psi_1] \\ I_1(k,y)=A(k,y)+2|E_0(k,y)||E_R(k,y)|\cos[\varphi(k,y)+\psi_0(k,y)+\Delta\psi_2] \\ \vdots \\ I_M(k,y)=A(k,y)+2|E_0(k,y)||E_R(k,y)|\cos[\varphi(k,y)+\psi_0(k,y)+\Delta\psi_M] \end{cases} \tag{2}$$

where E_O and E_R are the object beam and the reference beam, respectively, and $A=|E_0|^2+|E_R|^2$ is the zeroth order term. In Eq. (2), STDHs constructed by different sampling columns correspond to the tiny phase shifts, $\Delta\psi_m$ (m=1,..., M), added to the original information $\varphi(k,y)+\psi_0(k,y)$. In this case, the minimum phase shift obtainable is $\Delta\psi_{min}=(2\pi\Delta x\sin\theta)/p$, where p is the pitch of the spatial carrier, and θ is the angle between the fringe direction and the x-axis. If four proper columns, j,k,l,m, are selected, STDHs with $\Delta\psi_{j,k,l,m}=[0,\pi/2,\pi,3\pi/2]$ are obtainable, which is sufficient to apply a four-step phase shifting formula as in [1]. Then, numerical refocusing can be applied.

3 Experimental results

* Corresponding author: m.paturzo@isasi.cnr.it

https://doi.org/10.1051/epjconf/201921515001

We use a banded area of a CCD sensor to capture a set of observables while the object moves. For each object position, a column (or row) of pixels, which records a 1D signal, is acquired in the form of modulated fringes. Moreover, we use a diagonal pixel set to acquire 2D signals, which is an efficient way to scan that we recently introduced [6]. The acquired sequence is then used to synthetize the STDH that corresponds to the modulation of a new carrier. The idea of STDH is to move complexity from the reconstruction to the data capture process, by imposing a specific constraint on the scanning velocity that has to be matched to the frame rate of the sensor. Thus, a synthetic STDH can be created maintaining all the features of a space-space DH, along with some novel ones.

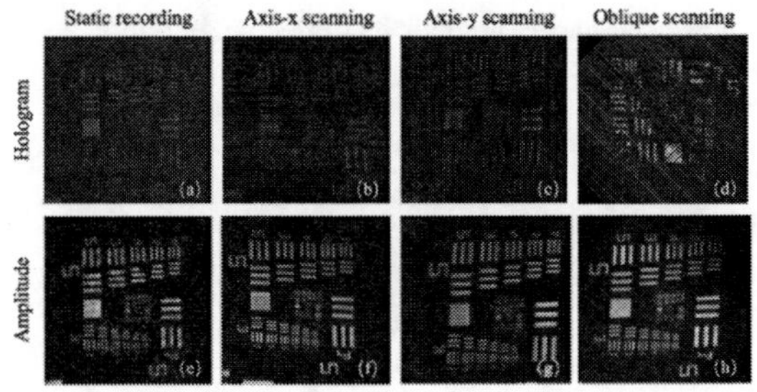

Fig. 1. Preliminary experimental results of self-assembling modality. (a) conventional hologram; (b)-(d) space-time holograms by x, y and oblique scanning; (e)-(h) amplitude reconstruction results.

To verify the resolution improvement process of self-assembling modality in STDH, a series of experiments were performed by using an USAF-1951 resolution target, as shown in Fig. 1. When the sample was scanned in 1D direction (x-axis or y-axis), the resolution under scanning direction could be improved by self-assembling of object frequencies [6]. Furthermore, once the sample was scanned in oblique direction, i.e. introducing velocity components both in x direction and y direction, the resolution of axis-x and axis-y could be improved at the same time. In summary, comparing STDH to conventional DH imaging, we notice that the object frequencies self-assembly process of space-time scanning can effectively improve the imaging resolution. Moreover, the Signal-to-Noise Ratio (SNR) of STDH is higher than the SNR of a conventional DH reconstruction, without requiring any further denoising process [7].

For onion cells, the holographic recording system is based on an off-axis Mach-Zehnder interferometer with coherent light source which wavelength is 532.8nm. The laser beam is divided into two beams by a polarizer beam splitter prism after simple filtering. The onion epidermal cells are placed on a 2D linear stage, it is used to achieve x-y plane scanning for STDH recording. Two 25× objective lenses are used for magnification imaging. A scientific-grade CCD camera is placed after beam splitter prism, the pixel size is 4.5μm × 4.5μm, and the screen size is 1280 × 960.

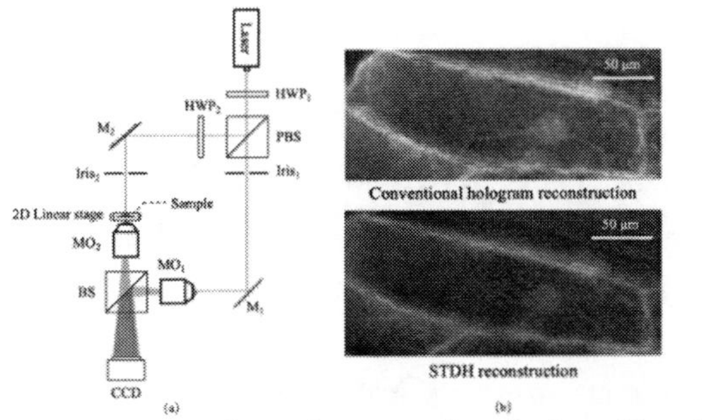

Fig. 2. STDH recording and reconstruction of onion epidermal cells. (a) experimental setup; (b) phase reconstruction results, the upper image is from conventional hologram, the lower image is from STDH.

As shown in Fig. 2(b), the phase image of STDH reconstruction results exhibits a clear resolution increase, with respect to conventional hologram reconstruction result. The internal structure of the onion epidermal cells can be observed more clearly. This is a further evidence that the self-assembly process improves resolution in 2D by 1D scanning imaging.

4 Conclusions

Thanks to the self-assembling of object frequencies, the resolution of Space-Time Digital Holography has been improved. In the preliminary experiment, we verified the improvement of imaging resolution by different scanning directions and proved that the imaging resolution in X and Y directions can be improved simultaneously in the oblique recording process. Also, in the experiment for onion epidermal cells, self-assembly process in STDH approach provided more detailed information in phase image with respect to conventional digital holography.

References

[1] V. Bianco, M. Paturzo, and P. Ferraro, Opt. Exp. **22**(19), (2014) ;
[2] V. Bianco, M. Paturzo, V. Marchesano, I. Gallotta, E. Di Schiavib and P. Ferraro, Lab Chip **15**(9), (2015) ;
[3] Biagio Mandracchia, Vittorio Bianco, Zhe Wang, Martina Mugnano, Alessia Bramanti, Melania Paturzo and Pietro Ferraro, Lab Chip **17**, 2831-2838 (2017) ;
[4] P. Memmolo, V. Bianco, M. Paturzo, P. Ferraro, Proc IEEE **105**(5), 892-905 (2017) ;
[5] M. K. Kim, SPIE Rev. **1**(1), 018005, 1-50 (2010) ;
[6] V. Bianco, Z. Wang, Y. Cui, M. Paturzo, and P. Ferraro, Opt. Lett. **43**(17), 4248-4251 (2018) ;
[7] V. Bianco, P. Memmolo, M. Leo, S. Montresor, C. Distante, M. Paturzo, P. Picart, B. Javidi and P. Ferraro, Light : Science & Applications **7**(1), 48 (2018).

Integration of fluidic channels and total internal reflection excitation with supercritical angle fluorescence detection

Kirsi Tappura[1,*], *Aki* Mäyrä[2], *Markku* Alamäki[2], *Annukka* Kokkonen[2], *Sanna* Aikio[2], and *Marianne* Hiltunen[2]

[1]VTT Technical Research Centre of Finland Ltd, FI-33101 Tampere, Finland
[2]VTT Technical Research Centre of Finland Ltd, FI-90571 Oulu, Finland

Abstract. In many diagnostic applications, there is a growing demand for small cost effective portable devices that enable fast on-site analysis. For this target, we are developing an integrated approach of a previously reported surface-sensitive measurement system based on total internal reflection excitation and supercritical angle fluorescence detection. The integrated measurement platform includes a polystyrene substrate on which the fluidic channels for sample transport, waveguides for fluorescence excitation and the readout optics are fabricated. In addition, thin layers with dielectric properties differing from those of polystyrene may be needed on the fluidic channels for optimal sample transport or biomolecule immobilisation. Here, special attention is paid to the influence of the dielectric environment near the solid-liquid interface on the fluorescence distribution and to the design of the readout optics. The first version of the designed readout optics of the integrated measurement platform is presented.

1 Introduction

Optical biosensors have attracted considerable research interest during the past several decades due to their versatile implementation possibilities, potential selectivity and immunity to electromagnetic interferences. Especially, fluorescence based detection has been considered beneficial with respect to sensitivity and achievable limits of detection. In many diagnostic applications, there is also a need for fast on-site analysis requiring small, cost effective portable systems. In this paper, a step towards this direction is taken by designing a low-cost polymeric substrate that enables both the fabrication of the fluidic channels for optimal sample transport and immobilisation of biomolecules combined with waveguide based total internal reflection (TIR) excitation and optical readout utilizing supercritical angle fluorescence detection (SAF) [1]. In the proposed approach, the SAF emitted by the molecules bound to the bottom of the fluidic channel generate a specific radiation distribution that is defined by the dielectric properties of the sample liquid and the substrate as well as by the distance of the fluorescing molecule from the bottom of the fluid channel [2-3]. Here, different dielectric environments near the bottom of the fluidic channel are studied in relation to the SAF fluorescence emission distribution that is then used as an input for designing the readout optics of the integrated measurement system described below.

2 Materials and methods

2.1. SAF emission

The SAF emission distributions are calculated for a randomly oriented dipole located at different distances from the solid-liquid interface at the bottom of the fluidic channel following the presentation of Enderlein et al. [3]. The results are calculated for two different liquid samples, water and milk, with the emission wavelength of the fluorescence being 670 nm (Figure 1).

Table 1. Optical parameters used in the simulations.

	Polystyrene	Al_2O_3	Water	Milk
Refractive index	1.58645	1.76581	1.33	1.344

2.1. Optics

In the integrated measurement platform (Figure 2), the optical readout device combines the TIR based excitation channel [1] and SAF based parabolic readout optics as presented in ref. [2]. In the excitation light guide, only the angles above the critical angle are allowed and can be controlled while coupling the light into the light guide.

The fluorescent emission is collected by a parabolic collimator lens, which has the focal point at the SAF emission point. A cylindrical aperture blocker and high-

* Corresponding author: kirsi.tappura@vtt.fi

pass filters limit the angular and spectral band before the light is focused to a detector by a plano-convex lens.

The integration sets some restrictions to the size and, thus, to the collection efficiency of the parabolic concentrator. First, the minimum diameter of the parabolic lens at the narrow end is limited by the space reservation of the fluidic channel and light guide. Secondly, the thickness of the whole system is limited by the thickness reservation of the excitation channel and the system thickness in general. Especially the overlap of the relatively wide excitation channel used in this first prototype and the parabolic concentrator reduces the collection efficiency.

Polystyrene (PS) was selected as the material for the optical readout device. Its high refractive index, when compared to other plastics, together with the possibility to use injection moulding, were the main reasons for the material selection.

The readout device was designed using the Optics Studio non-sequential design tool. The simulated angular distributions in polystyrene were modelled as a pointwise defined radial source.

3 Results and discussion

3.1 SAF emission distributions

Figure 1 shows the calculated fluorescence distributions expressed as normalized radiated power as a function of the emission angle in the polystyrene substrate. The results are shown for water and milk as the sample matrices with and without an Al_2O_3 coating of 5 nm in thickness on the bottom of the fluid channel. Due to their fairly similar refractive indices, the curves for water and milk follow closely each other, but the 5 nm thick Al_2O_3 coating has a significant effect especially at high emission angles. The highest power is obtained at the critical angle.

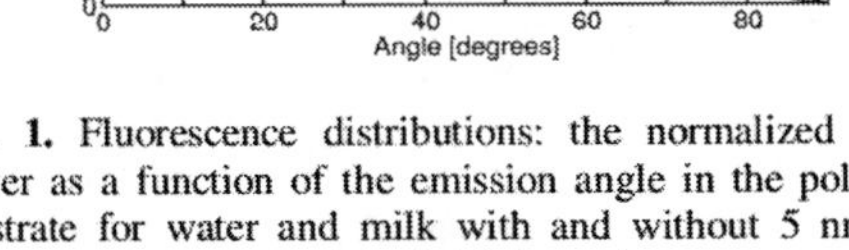

Fig. 1. Fluorescence distributions: the normalized radiated power as a function of the emission angle in the polystyrene substrate for water and milk with and without 5 nm Al_2O_3 coating on the bottom of the fluid channel.

3.2 Optics

The optical system design and spatial irradiance distribution of the reflected rays are shown in Figure 2. The fluidic channel locates on the light guide on the surface opposite to the readout optics although not shown in the figure. The diameter of the collection optics is 12.5 mm and the thickness of the parabolic concentrator 3.5 mm. The parabolic concentrator can collect SAF angles at about 57.5 degrees and above. The irradiance profile visualises how the irradiance increases towards the edges of the spatial filter that is towards the critical angle. When reflection losses are excluded, the total SAF collection efficiency is about 30% and it is mainly limited by the overlap of the excitation channel and parabolic concentrator. The system will be injection moulded and it is assumed that some modifications has to be made in order to optimise the system for the specific applications.

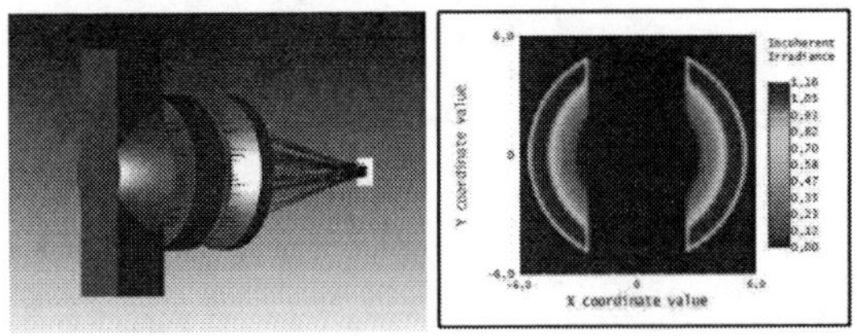

Fig. 2. The measurement platform showing the (vertical) light guide and the readout optics with the parabolic lens (left) and the relative irradiance distribution of the reflected rays at the spatial filter to be focused to the detector (right).

3 Conclusions

It was shown that even very thin additional layers on the fluidic channels can have a significant influence on the SAF emission distribution that need to be taken into account when designing the readout optics.

The readout optics simulations show that the SAF angles higher than 57.5 degrees can be measured with the integrated measurement platform and that the overall SAF collection efficiency is about 30% for the relatively wide excitation channel used in the current design. The collection efficiency is mainly limited by the overlapping parabolic lens and excitation channel, in addition to the inherent property of the SAF detection rejecting the emission in the angles below the total internal reflection.

The SAFTIR project ("Point-of-need diagnostics for animal health and food safety") is funded by Business Finland and VTT Technical Research Centre of Finland Ltd.

References

1. K. Tappura, H. Välimäki, US 10018563 B2.
2. H. Välimäki, K. Tappura, Chem. Phys. Lett. **473**, 358 (2009)
3. J. Enderlein, T. Ruckstuhl, S. Seeger, Appl. Optics **38** 724 (1999)

 https://doi.org/10.1051/epjconf/201921515003

Addressing the Fabrication Difficulties of Femtosecond Laser Written Surface Waveguides for Enhanced Evanescent Coupling

Vítor A. Amorim[1,2,*], *João M.* Maia[1,2], *Duarte* Viveiros[1,2], and *P. V. S.* Marques[1,2]

[1]Centre for Applied Photonics, INESC TEC, Porto 4150-179, Portugal
[2]Department of Physics and Astronomy, University of Porto, Porto 4169-007, Portugal

Abstract. In this work, the fabrication of optical waveguides embedded in fused silica (Suprasil1) and boro-aluminosilicate glass (Eagle2000) is demonstrated with femtosecond laser direct writing, as well as their suitability to be brought to the surface, through wet etching, for enhanced evanescent coupling with the external dielectric medium. Fused silica demonstrated to be inappropriate in this particular application, as the guiding region is at the bottom of the induced modification, creating a barrier between the guided mode and the substrate's boundary. Furthermore, the existence of nanogratings meant that, upon contact of the top of the induced modification with the substrate's boundary, the waveguide is quickly etched. Eagle2000 demonstrated to be superior to fused silica due to its characteristic modification cross-section and absence of nanogratings, which allowed the placement of the guiding region as close to the substrate's surface as required. However, surface roughness arising from the creation of insoluble products in the HF solution was found. The addition of HCl to dissolve these products was implemented.

1 Introduction

Femtosecond laser direct writing is a fabrication technique that relies on a very localized modification to manufacture three-dimensional structures. In several materials, the modification translates into an increased refractive index, enabling the fabrication of optical circuits [1]. The creation of nanometric structures aligned along a preferential direction (nanogratings) is also possible in fused silica, leading to an anisotropic etching reaction that enables the fabrication of microfluidic systems and, when in conjunction with optical circuits, optofluidic devices [2].
Being an inherently three-dimensional technique, it allows the fabrication of waveguides at any depth, including near the surface which is of great interest in sensing applications. However, ablation sets a limit on how shallow a waveguide can be. In this work, we employ the femtosecond laser direct writing technique in the fabrication of optical waveguides at low depths in Suprasil 1 and Eagle2000 substrates, and discuss the outcome of etching them to the substrate's surface.

2 Experimental Procedure

A fiber amplified femtosecond laser, with a second harmonic beam at 515 nm and a pulse duration of approximately 250 fs, was used to induce the refractive index modifications. These were formed by transverse scanning, with the linearly polarized beam focused at a depth of 50 μm via a 0.55 numerical aperture aspherical lens. Laser beam polarization was set parallel to the writing direction, and all waveguides were written on the same scanning direction to avoid the Quill effect.
Wet etching was made in an ultrasonic bath, with the side facets of the glass substrates being polished after the etch. To determine the spectral characteristics of the waveguides, light was coupled into a single-mode fiber (SMF-28) and butt-coupled to the entrance facet of the modification tracks. The light collected by a second butt-coupled optical fiber at the exit facet was inspected in an optical spectrum analyzer.

3 Experimental Results and Discussion

For this study, an array of optical waveguides was fabricated in fused silica (Suprasil 1) while increasing the inscription depth in steps of 0.5 μm. As can be seen in Fig. 1 (a), the induced modifications' cross-section has an upper region composed by a decreased refractive index followed by a lower guiding region. To expose the guided mode to the external dielectric medium, the uppermost region must be etched. However, the presence of nanogratings translate into the etch of the laser induced modification as soon as the top of the modified region contacts the substrate's surface.
An alternative is to replace the fused silica with boro-aluminosilicate glass (Eagle2000). In addition to the nonexistence of nanogratings in the laser induced modification in Eagle2000, the upper region composed by a decreased refractive index is also absent, as can be seen in Fig. 1 (b). This means that the guiding region, and consequently the guided mode, can be brought as close to the substrate's surface as desired, enabling the capability of tuning the evanescent field coupling to the exterior.

* Corresponding author: vitor.a.amorim@inesctec.pt

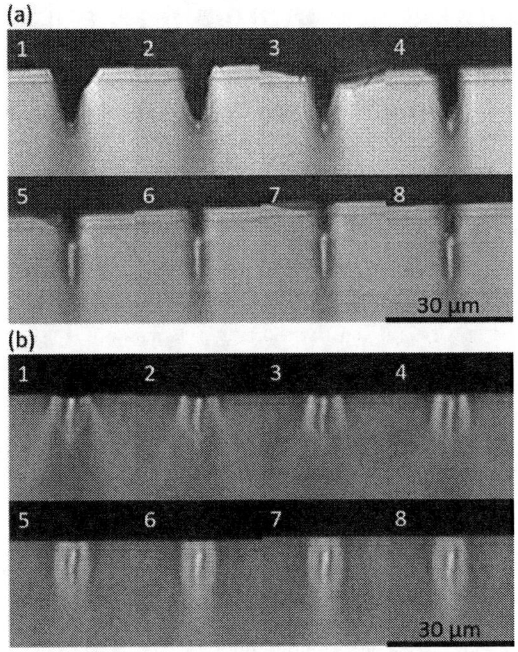

Fig. 1. Cross-section of waveguides fabricated in Suprasil 1 (a) and Eagle 2000 (b) at a depth of $50 + 0.5 \times n$ µm (n=waveguide number) after being etched to the surface in HF 10%.

Another advantage of Eagle2000 is the lower Rayleigh scattering induced in femtosecond laser written optical waveguides when compared with fused silica [3], enabling low-loss broadband light guidance, as can be seen in Fig. 2. Their transmission characteristics also depend on the amount of etched waveguide, displaying high insertion losses at long wavelengths due to the reduced core dimensions and higher insertion losses at short wavelengths due to reduced coupling efficiency and increased Rayleigh scattering from nanometric defects at the substrate's surface.

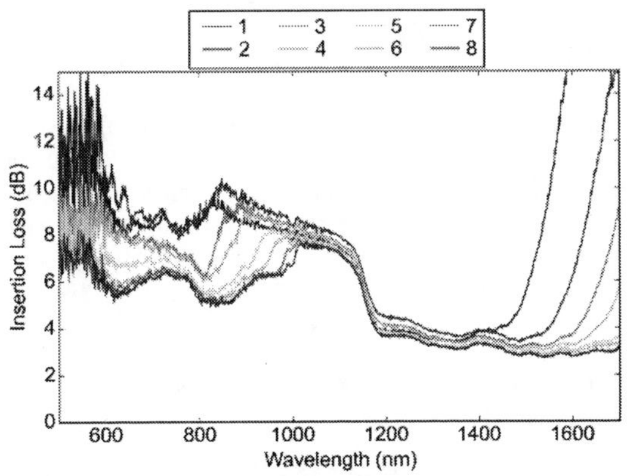

Fig. 2. Insertion loss of 2.4 cm long waveguides fabricated in Eagle 2000 at a depth of $50 + 0.5 \times n$ µm (where n is the waveguide number) after being etched to the surface in HF 10%.

Fig. 3 displays the surface topography of Eagle2000 before and after etching with HF 10%. As can be seen, the surface roughness increases significantly after etching, limiting possible applications such as the deposition of metallic thin-films for plasmonic sensors. The roughness is connected with the creation of insoluble products in the HF solution, which deposit on the generated surface and act as a masking layer [4]. By adding HCl the products become soluble, improving the etched surface's quality, as visible in Fig. 3 for a solution of HF 1%/HCl 37% for a volume ratio of 10/1.

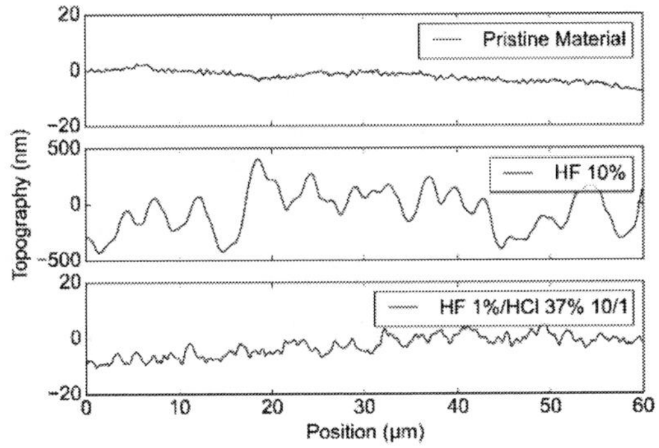

Fig. 3. Surface topography of Eagle2000 with a pristine surface, after being etched in HF 10%, and after being etched in HF 1%/HCl 37% for a volume ratio of 10/1.

4 Conclusions

Eagle2000 has shown to be suitable for the fabrication of surface waveguides using femtosecond laser direct writing and wet etching. The nonexistence of nanogratings in the laser induced modification and the absence of an upper region composed by a decreased refractive index, as in fused silica, enables the placement of the guiding region as close to the substrate's surface as desired. Furthermore, waveguides fabricated in this glass present low-loss broadband guiding characteristics, suiting a broad range of applications. Increased surface roughness, created by insoluble products when etching with HF, was identified, being minimized with the application of HCl in the etchant solution.

This work was supported by Fundação para a Ciência e a Tecnologia through grant no. SFRH/BD/128795/2017 and by project "NanoSTIMA: Macro-to-Nano Human Sensing: Towards Integrated Multimodal Health Monitoring and Analytics/NORTE-01-0145-FEDER-000016", financed by the North Portugal Regional Operational Programme (NORTE 2020), under the PORTUGAL 2020 Partnership Agreement, and through the European Regional Development Fund (ERDF).

References

1. V. A. Amorim, J. M. Maia, D. Alexandre, P. V. S. Marques, J. Light. Technol. **35** 7 (2017)
2. J. M. Maia, V. A. Amorim, D. Alexandre, P. V. S. Marques, J. Light. Technol. **35** 8 (2017)
3. V. A. Amorim, J. M. Maia, D. Viveiros, P. V. S. Marques, J. Light. Technol. 6 (2019)
4. C. Iliescu, J. Jing, F. E. H. Tay, J. Miao, T. Sun, Surf. Coat. Tech. **198** 5 (2005)

Hybrid Organic-Inorganic photoresists, a promising class of materials for Optofluidic integration.

Jinane Elias[1,*], *Pascal* Etienne[1], *Sylvie* Calas-Etienne[1] and *Laurent* Duffours[2]

[1]Laboratoire Charles Coulomb (L2C), University of Montpellier, CNRS, Montpellier, France
[2]PRIME Verre, 1350 Albert Einstein, Montpellier, France

Abstract. Through the efforts to fuse planar optics and microfluidics in order to produce dye lasers, biosensors, trapping and cell sorting device, we can notice the rising interest in optofluidics since early and mid 2000's. However mass production of these devices heavily relies on fast and easy patterning of the constituent material. PDMS, being one of these materials, gained an added value because of its elasticity, hydrophobicity and permeability to gaz. Nonetheless, these specifications are not convenient for all types of applications. The growing capability to use Hybrid Organic-Inorganic materials for the fabrication of integrated optics components and microfluidic channels is what makes this class of materials an ideal candidate for this integration. This work aims to implement, on the same chip, an optical and a microfluidic layer using Sol-Gel processing of Organic-Inorganic materials. The interest in this vertical integration arises from the need to manipulate the fluid in the microchannels using evanescent field optical pressure.

1 Introduction

Optofluidics refer to a wide range of applications in which optics and fluidics are integrated on the same device and used synergistically to synthesize creative, novel and highly versatile functionalities, in which light is used to manipulate and analyse fluids or vice versa. Even though the integration of an optical component (light sources, filters or photodetectors) into the lab-on-a-chip has recently gained lots of interest, with the efforts to fuse planar optics and microfluidics, their commercialization remains greatly challenged by the ability to mass produce these devices while maintaining low cost hence disposability.
PDMS appeared to be the best candidate for mass production of these devices. However it presents some drawbacks, such as absorption of organic solvents, lipidic like and hydrophobic molecules and subsequent swelling of the chip, permeability to gas, and minor auto-fluorescence (in comparison with other polymers) in the Near UV-Vis-Near IR (240–1100 nm). These drawbacks limit its use to certain applications where these problems don't affect the final output. These problems, especially the hydrophobic nature of PDMS, have triggered many research projects in order to modify the surface of PDMS using special coatings. As an alternative, E. Mechref et al. have developed an organic-inorganic positive photoresist [1], suitable for microfluidic channels fabrication using single step (deposition, insulation, and engraving) sol-gel processing. This material can be tailored to be as good as PDMS and can still overcome its undesired properties. One problem arises, which is the control of fluids in these rigid micro channels and the implementation of the basic microfluidic components such as active mixers, valves and sensors. What we propose as a solution is evanescent field optical control through an integrated optical layer which was already implemented in previous works in our Lab[2] Our main purpose is to design and implement optofluidic control elements for lab on chip applications such as optofluidic mixer as a first step.

1.1 Hybrid Organic Inorganic Materials

1.1.1 Hybrid materials definition

As the name suggests they are a set of material in which two networks coexist: an organic and inorganic one, giving rise to a hybrid material that can be tailored, in a way to manifest the properties of both its constituents. The inorganic component increases hardness and resistance to elevated temperature while the organic network ensures good elasticity, tenacity, ductility which allows us to increase the thickness of the coatings without significant cracks. Precursors are ORganically MOdified CERamics having a general chemical formula of $R_xM(OR')_{4-x}$. The alkoxide group OR' is responsible of the mineral network elaborated by sol-gel process: a sol that turns into a gel after hydrolysis and condensation reactions. The R groups are responsible for the organic network elaborated through selective UV induced polymerisation or through thermally induced polymerization.Due to their unique and intermediate properties of the mineral and organic, these materials are of great interest for many applications in optics, and microfluidics.

1.1.2 Negative Tone Hybrid photoresists for integrated optics

Since 1992 the work of different groups has shown the effectiveness of a composition based on a mixture of methacryloxypropyltrimethoxysilane (MAPTMS) and zirconium propoxide ($Zr(OPr)_4$. The addition of $Zr(OPr)_4$ to the PMMA network is used to adjust its optical properties, notably to increase the refractive index. It also enhances the quality of the waveguide profile after

* Corresponding author: jinane.elias@umontpellier.fr

https://doi.org/10.1051/epjconf/201921516001

revelation because it leads to a well-polymerised and hydrophilic surface under UV-illumination. Subsequent studies have shown the potential of this system to yield an index contrast of 0.013 [2] with the addition of an adequate photoinitiator. It was shown that the higher the refractive index contrast, the more intense the evanescent field can be [3]. For this reason we chose to proceed with a mixture of MAPTMS and $Zr(OPr)_4$ while trying to increase the contrast by changing the photoinitiator and the $Zr(OPr)_4$ to MAPTMS proportion. In addition, this simple fabrication process of waveguides is very adequate to mass production.

1.1.3 Positive Tone Hybrid photoresists for microfluidic chips

A positive tone resist is a good alternative for negative tone photoresists when it comes to the processing of channels by laser writing. A hybrid material offers better command on the properties of the microfluidic channels because of the reasons cited in 1.1.1. This is why E. Mechref et al. [1] developed an organic-inorganic chemically amplified positive photoresist based on Polyamic acid (PAA) and suitable for microfluidic channels fabrication using single step (deposition, insulation, and engraving) sol-gel process. PAA is well known for its good mechanical properties and high thermal stability, enhanced even more through the addition of 4-vinylether-phenyltriethoxysilane (VEPTES) pre-hydrolyzed by sol-gel method to form the inorganic network.

1.2 Evanescent Field Optical Control

Light carries energy and momentum; optical forces result from the transmission of this energy to another system. On the macroscopic scale, these forces pass unnoticed. However, the acceleration induced by a light beam with a moderate intensity, on relatively small particles is effective to move, trap, or guide a particle in the Mie or the Rayleigh regime. This became experimentally feasible after the invention of lasers and has been subject to implementation since early 1970's in the work of Ashkin [4]. Kawata and Sugiura [5]reported that particles with diameters of 1–27 μm present within the penetration depth of the evanescent field can be attracted to the core by the gradient force and propelled along a single mode fiber by the scattering force in the same direction as the mode propagates. Theoretical examination of optical force exerted on a dielectric sphere by the Gaussian beam the evanescent field was also conducted.

2. Results

Initial studies have demonstrated the feasibility of single mode waveguides written due to selective photo-induced increase of refractive index in a hybrid organic-inorganic negative type photo resist, composed of hydrolysed MAPTMS and $ZrPO_4$ chelated by MAAH and further hydrolysed. After UV curing it showed an index rise of around 0.002 in the visible. Refractive index measurements were made using an ATAGO's Multi-Wavelength Abbe Refractometer DR-M2 having an instrumental error of$\pm 5 \times 10^{-5}$. Unfortunately the statistical error appears to be of the same order of the difference which makes the significance of these results doubtable. Another measurement method is under investigation. Light confinement in these waveguides was confirmed. Attenuation in the visible range (at 625nm) will be quantified, alongside with the attenuation due to bending. Other mixtures such as (3 Glycidyloxypropyl)trimethoxysilane (GPTMS) were investigated; it showed on the other hand a decrease in the refractive index due to insulation.

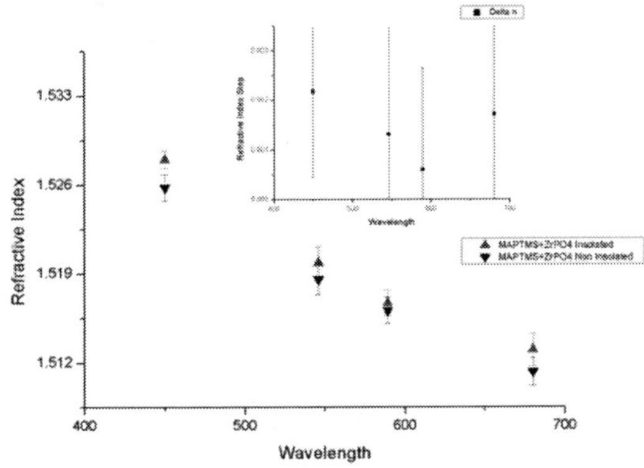

Figure 1: Refractive Index of cured and non cured MAPTMS (annealed at 110 degrees Celsius) and refractive index step (Inset)

3. Conclusion

As a conclusion, hybrid organic-inorganic materials are biocompatible, mechanically robust and impermeable materials suitable for mass production of optofluidic Lab on Chips. Waveguides can be easily written by photopolymerisation of negative photoresists, and microfluidic channels revealed in positive resist after laser writing. All of which makes it a low cost, high yield process.

References

[1] E. Mechref, J. Jabbour, S. Calas-Etienne, K. Amro, A.Mehdi, R. Tauk, D. Zaouk , P. Etienne, RSC Adv., **6** (2016)

[2] P. Coudray, P. Etienne, Y. Moreau, J. Porque, S. I. Najafi, Opt. Commun., **143** (1997)

[3] R. Bernini, N. Cennamo, A. Minardo, and L. Zeni, , IEEE Sens. J.,**6** (2006)

[4] A. Ashkin, Phys. Rev. Lett., **24** (1970)

[5] S. Kawata and T. Sugiura, Opt. Lett., **17** (1992)

https://doi.org/10.1051/epjconf/201921516002

Optical transport of fluorescent diamond particles inside a tapered capillary

Christophe Pin[1], *Ryohei* Otsuka[1], *Hideki* Fujiwara[1], *Keiji* Sasaki[1,*]

[1]Research Institute for Electronic Science, Hokkaido University, Kita 20-jo Nishi 10-chome, Kita-ku, Sapporo, Hokkaido, 001-0020, Japan

Abstract. Optical forces provide an efficient way to sort particles and biological materials according to their optical properties. However, both enhanced optical forces and a large interaction volume are needed in order to optically sort a large number of nanoparticles. We investigate the use of a tapered glass capillary as an optofluidic platform for optical manipulation and optical sorting applications. Tapered capillaries with micrometre and sub-micrometre sizes are fabricated. After filling the tapered capillary with a colloidal solution of red fluorescent diamond particles, a green laser light is coupled into the capillary. The tapered capillary acts both as a microfluidic channel and as an optical waveguide, making it possible for the light to interact with the particles inside the sample solution. Using an incident laser power of few tens of milliwatts, we achieve optical transportation of the brightest particles inside the tapered part of the capillary. Particle velocities as high as few tens of micrometres per second are measured.

1 Introduction

Based on optical forces, optical sorting provides an efficient and contactless method to separate materials in liquid suspension according to their optical properties [1-8]. Various properties of light such as its wavelength, its polarization, and its spatial distribution, can be used to tailor the optical forces being applied to different particles in solution. As light-matter interactions depend not only on the light properties but also on the optical properties of materials, selective optical trapping or transportation can be achieved and used in order to optically sort nanoparticles with different optical properties. Such an optical sieve is highly needed for sorting a large variety of nanomaterials. However, efficient optical manipulation of a large number of nanoparticles remains a challenging task since it would require both an intense laser light to cover a large interaction volume, and near-field enhanced optical forces that drastically reduce the size of the interaction volume [4,9-15].

In order to overcome this issue, we focus here on the fabrication and the use of tapered glass capillaries for optofluidic applications such as nanoparticle optical manipulation. First, we present numerical simulation results in order to evaluate the optical properties of glass micro- and nanocappillaries. In a second part, we describe how micrometre and sub-micrometre size glass capillaries can be fabricated. Eventually, we report on the optical transport of fluorescent diamond particles inside a tapered capillary.

2 Results and discussion

1.1 Micro- and nanocapillary waveguides

We consider here fused-silica glass micro- and nano-capillaries filled with water solution and surrounded by air. The refractive index of both capillary materials (n_{glass} = 1.46, n_{water} = 1.33) being higher than the refractive index of air (n_{air} = 1.0), the tapered capillary acts as an optical waveguide with a partially liquid part. We numerically investigate the optical guiding properties of micro- and nanocappillaries in order to evaluate the overlap between the guided light and the liquid channel at different wavelengths and for different capillary sizes. Figure 1 shows the electric field distributions of the fundamental guided mode that can be excited in water-filled capillaries with different diameters at 1064 nm and 532 nm wavelengths.

Although reaching the single-mode regime requires tapered capillaries to have sub-micrometre dimensions, it is found that such nanocappillaries can offer a large overlap between the guided light and le liquid solution. Larger multimode capillaries may still provide an optofluidic platform for studying light-matter interactions as long as a non-negligible part of the guided light propagates in the evanescent field near the glass-water interface. In such a case, the light propagation in the tapered capillary is expected to depend on the light coupling condition at the entrance of the capillary, as well as on the taper geometry.

* Corresponding author: sasaki@es.hokudai.ac.jp

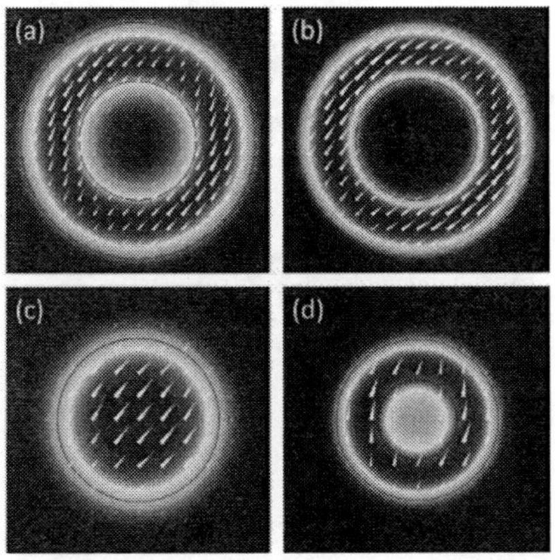

Fig. 1. Cross-sections of the electric field distribution of the fundamental guided mode (a),(b) in a 2 µm large capillary and (c),(d) in a 700 nm large capillary at (a),(c) 1064 nm and (b),(d) 532 nm wavelengths.

1.2 Fabrication of liquid-filled tapered capillaries

We investigated the fabrication of tapered glass capillaries (as shown in Figure 2) using a standard heating and pulling process originally developed in order to fabricate tapered optical fibres. Figure 3 shows the pulling-time dependence of the waist outer diameter of the fabricated tapered capillaries. Tapered capillaries with sub-micrometre diameters down to 200 nm were successfully fabricated after optimization of the pulling time and speed.

A home-made system was developed in order to inject a liquid sample solution inside a tapered capillary. For the experimental demonstration of particle optical transport inside a tapered capillary, fluorescent diamond particles were chosen for their high refractive index value (2.44 at 532 nm) and their bright and robust fluorescence due to the presence of NV- centers (no photobleaching or blinking effects) [16]. We used a water solution containing red fluorescent diamond particles with an average size of 100 nm that was purchased from FND Biotech.

After injecting the liquid solution inside the capillary, both ends of the capillary were cleaved, and green laser light (at 532 nm wavelength) was coupled to one extremity of the capillary. We used a cleaved optical fibre coated with immersion oil to improve the coupling efficiency. Observation and recording of the particles' motion inside the capillary was done using a standard fluorescence microscope.

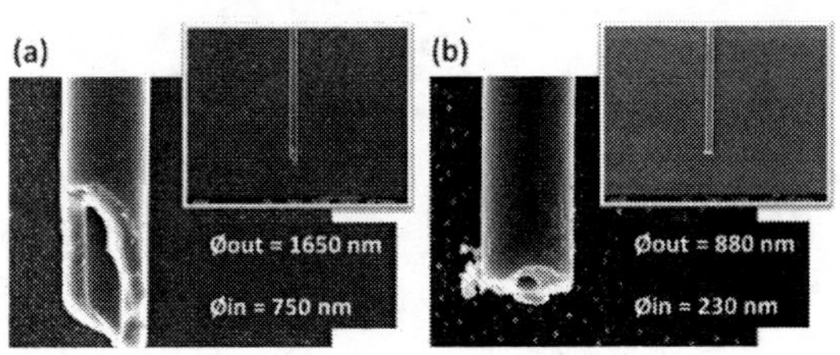

Fig. 2. Scanning electron microscopy images of two broken tapered glass capillaries of different sizes.

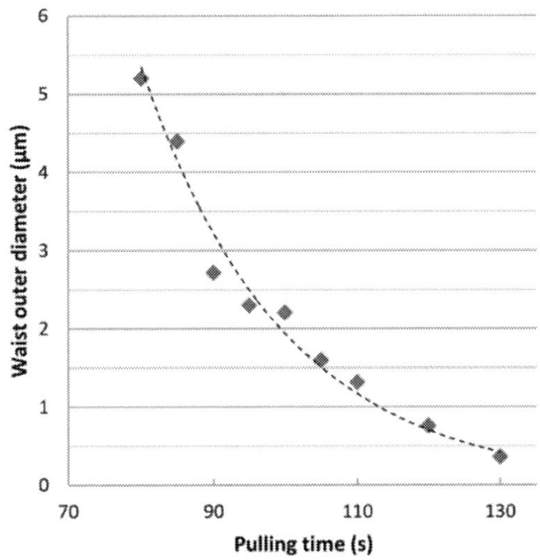

Fig. 3. Waist outer diameter of the fabricated tapered capillaries as a function of the pulling time (pulling speed: 340 µm/s).

1.3 Optical transport of fluorescent diamonds

During the experiment, a transmission efficiency reaching about 20% was measured. This value takes into account both the fibre-to-capillary coupling losses and the radiative losses occurring in the tapered part of the capillary. Using a tapered capillary of 6.7 µm outer diameter, the brightest diamond particles were successfully propelled inside the capillary when the laser was switched on. We used an input laser power of few tens of milliwatts. Similar results were obtained using a tapered capillary with a 2.2 µm outer diameter. The recorded particles' motion was then analysed, and particle velocities reaching few tens of micrometres per second were measured. We were able to control the particle velocity by varying the input laser power.

3 Conclusion

In this work, we investigated the fabrication and use of tapered glass capillaries for nanoparticle optical transport and sorting applications. We first studied the optical properties of tapered capillaries using a numerical model. Then we fabricated tapered glass capillaries filled with a colloidal solution of red fluorescent diamond particles, and we performed optical transport experiments using a green laser light coupled into the capillary waveguide. The results presented here are expected to pave the way for further investigation on the use of tapered glass capillaries as an optofluidic platform for optical sorting and other applications.

We acknowledge financial support from JSPS KAKENHI (Grant Numbers JP16H06506, JP18H05205, and JP18H03882).

References

1. T. Imasaka, Y. Kawabata, T. Kaneta, Y. Ishidzu, Anal. Chem. **67**, 1763 (1995)
2. S.J. Hart, A.V. Terray, Appl. Phys. Lett. **83**, 5316 (2003)
3. M.P. MacDonald, G.C. Spalding, K. Dholakia, Nature **426**, 421 (2003)
4. M. Ploschner, T. Cizmar, M. Mazilu, A. Di Falco, K. Dholakia, Nano Lett. **12**, 1923 (2012)
5. G. Tkachenko, E. Brasselet, Nat. Commun. **5**, 3577 (2014)
6. S.E.S. Spesyvtseva, S. Shoji, S. Kawata, Phys. Rev. Appl. **3**, 044003 (2015)
7. F. Nan, Z. Yan, Nano Lett. **18**, 7400 (2018)
8. Y. Shi, S. Xiong, L.K. Chin, J. Zhang, W. Ser, J. Wu, T. Chen, Z. Yang, Y. Hao, B. Liedberg, P.H. Yap, D.P. Tsai, C.-W. Qiu, A.Q. Liu, Sci. Adv. **4**, eaao0773 (2018)
9. A.H. Yang, S.D. Moore, B.S. Schmidt, M. Klug, M. Lipson, D. Erickson, Nature **457**, 71 (2009)
10. S.E. Skelton, M. Sergides, R. Patel, E. Karczewska, O.M. Marago, P.H. Jones, J. Quant. Spectrosc. Radiat. Transf. **113**, 2512 (2012)
11. A. Cuche, A. Canaguier-Durand, E. Devaux, J.A. Hutchison, C. Genet, T.W. Ebbesen, Nano Lett. **13**, 4230 (2013)
12. P. Schein, D. O'Dell, D. Erickson, ACS Photon. **4**, 106 (2017)
13. M. Soltani, J. Lin, R.A. Forties, J.T. Inman, S.N. Saraf, R.M. Fulbright, M. Lipson, M.D. Wang, Nature Nanotech. **9**, 448 (2014)
14. J.E. Baker, R.P. Badman, M.D. Wang, WIREs Nanomed. Nanobiotechnol. **10**, e1477 (2018)
15. C. Pin, J.-B. Jager, M. Tardif, E. Picard, E. Hadji, F. de Fornel, B. Cluzel, Lab. Chip **18**, 1750 (2018)
16. W.W.W. Hsiao, Y.Y. Hui, P.C. Tsai, H.C. Chang, Acc. Chem. Res. **49**, 400 (2016)

AUTHOR INDEX

AUTHOR INDEX

AUTHOR INDEX

EDP Sciences – Web of Conferences
17, Avenue du Hoggar
Parc d'Activité de Courtabouf
BP 112
F-91944 Les Ulis Cedex A
France

ISBN 978-1-5108-9425-9

CURRAN ASSOCIATES INC.
proceedings.com

International Conference "Mathematical Modelling in Biomedicine" 2019

ITM Web of Conferences Volume 31 (2020)

Moscow, Russia
30 September – 4 October 2019

edp sciences